T0256962

States of Disease

POLITICAL ENVIRONMENTS
AND HUMAN HEALTH

Brian King

UNIVERSITY OF CALIFORNIA PRESS

University of California Press, one of the most distinguished university presses in the United States, enriches lives around the world by advancing scholarship in the humanities, social sciences, and natural sciences. Its activities are supported by the UC Press Foundation and by philanthropic contributions from individuals and institutions. For more information, visit www.ucpress.edu.

University of California Press
Oakland, California

© 2017 by The Regents of the University of California

Library of Congress Cataloging-in-Publication Data

Names: King, Brian (Brian Hastings), 1973– author.
Title: States of disease : political environments and human health / Brian King.
Description: Oakland, California : University of California Press, [2017] | Includes bibliographical references and index.
Identifiers: LCCN 2016038478 (print) | LCCN 2016039497 (ebook) | ISBN 9780520278202 (cloth : alk. paper) | ISBN 9780520278219 (pbk. : alk. paper) | ISBN 9780520962118 (epub)
Subjects: LCSH: Social ecology—Health aspects—South Africa. | Social ecology—Health aspects—Botswana. | AIDS (Disease)—Treatment—Government policy—South Africa. | AIDS (Disease)—Prevention—Government policy—South Africa. | Environmental health—Botswana.
Classification: LCC HM861 .K56 2017 (print) | LCC HM861 (ebook) | DDC 304.2/0968—dc23
LC record available at https://lccn.loc.gov/2016038478

Manufactured in the United States of America

25 24 23 22 21 20 19 18 17
10 9 8 7 6 5 4 3 2 1

States of Disease

To Cliff Shikwambane
  The world is less bright without you

And to Erica King
  For everything

Contents

Figures

Preface

Reading the *New York Times* on December 2, 2009, I was immediately drawn to two separate stories highlighted on the front page. The first, "New hopes on health care for American Indians," chronicled a meeting between the leaders of 564 American Indian tribes and President Barack Obama and members of his cabinet. The meeting was designed to highlight the inequities within the national health-care system that resulted in disease vulnerabilities and constrained access to treatment options in indigenous communities. Noting that American Indians die of infectious diseases and illnesses—including tuberculosis, alcoholism, diabetes, and pneumonia—at higher rates than other members of the national population, President Obama emphasized the importance of doing more to "address disparities in health care delivery."[1] While notable for its historical significance, this was clearly a strategic element of the administration's push for health-care reform that culminated in the passage of the Affordable Care Act in 2010. In detailing the meeting, the article highlighted the multiple factors complicating the provision of services on Indian reservations and the continued sentiment of some residents that historical land-dispossession has contributed to the underlying structural conditions producing health within the reservation system. Roughly one-third of American Indians were uninsured, and a quarter lived below the national poverty level. Additionally, the lack of needed medical equipment in local health-care facilities, and often long travel distances to access services, further intensified the constraints to care-seeking for conditions that could become life-threatening. Senator Byron L. Dorgan of North Dakota, who was interviewed for the article, commented that health-care rationing was occurring on Indian reservations and that their citizens were "'first Americans' living in third world conditions."

The second story, "Breaking with past, South Africa issues broad AIDS policy," chronicled the efforts being taken by South African President Jacob Zuma in moving the national government to respond more aggressively to provide treatment for HIV/AIDS. In a speech corresponding with World AIDS Day, President Zuma pledged to treat HIV-positive individuals with tuberculosis earlier, in accordance with guidelines from the World Health Organization. Tuberculosis is the leading cause of death for HIV-infected South Africans, and the declaration was hailed as a response that would save lives. In the speech, Zuma urged South Africans to learn their HIV status and use "condoms consistently and correctly during every sexual encounter."[2] The speech was widely praised, and the American ambassador to South Africa quickly responded by pledging an additional $120 million in funding over two years to help in the provision of antiretroviral drugs (ARVs) for those living with HIV.

Yet, enmeshed in these assertive declarations and optimistic responses from members of the international community, the article emphasized how these actions should be seen as a departure from the policies of the previous administration led by the former president, Thabo Mbeki. The story highlighted statements made by some governmental officials during Mbeki's tenure that doubted the effectiveness of ARVs as a course of treatment for HIV, which contributed to delays in their provision to South African citizens. It was noted that Mbeki had previously questioned the link between HIV and AIDS and once remarked that he had never known anyone who had died from the disease. Two days before the publication of the *New York Times* article, the Congress of South African Trade Unions declared that Mbeki should formally apologize for failing to fight an epidemic that had destroyed more lives than an invading army. Not to be outdone, the Young Communist League insisted that Mbeki be prosecuted for genocide. The hope expressed by President Zuma's declarations, therefore, was tempered by the country's recent controversies over competing understandings of the HIV/AIDS epidemic. The more aggressive interventions by the national government continued to be interpreted through a filter that raised uncertainty about its responsibilities, along with the roles of public health institutions and the global community, to effectively respond to South Africa's HIV/AIDS epidemic.

While it would be possible to read these two newspaper articles as separate narratives about human health, I was struck by the coincidence of their coverage and points of convergence. Positioned as two disparate health circumstances in two different settings, the stories reinforce a tendency in scholarship and policy to locate illness at the site of the individual body while identifying supposedly distinct disease patterns. Yet, when they are placed next to each other, several themes are brought into stark relief. In each case, underlying structural conditions, both historical and spatial, are identified as having produced both disease and the opportunities for healthy decision making. Whether it is the inability to locate medications to address diabetes on a Navajo reservation in Arizona, or accessing antiretroviral therapy for HIV in South Africa, the possibilities for health and well-being are presented as having been constructed out of social and spatial relationships that had unfolded over time, thereby making certain members of a population more vulnerable to illness. Similarly, these two cases say much about the complicated entanglements between the state and the body politic. Both move seamlessly between the potentially contradictory positions of individual responsibility and accountability of the national government in creating the conditions for the spread of disease. While simultaneously locating disease at the site of the individual body, the role of the state in providing either information or medications is assumed and seen as instrumental in the historical creation of illness or the production of health in the contemporary period. Finally, these cases suggest that the ways in which human health is understood, and also contested, are inscribed upon a landscape that involves multiple actors, social networks, and relational exchanges. These entanglements include negotiations over access to crucial services or medications but also involve competing discourses of disease and well-being. The interactions between these factors shape the lived experiences of individuals while producing vulnerabilities to disease and the possibilities for human health.

These two stories also leave much unsaid. In describing human vulnerabilities to the spread of disease, they fail to provide complete accounts of the social ecology of health in these different settings. While distance to health-care facilities and the ability to access medications are critically important in responding to disease, much more is at stake. Human health is not the absence of disease but a fluid domain that includes the ability to realize aspirations and have the capacity to make decisions for one's own mental and bodily state. It is a human right that is achieved when there is physical, mental, and social well-being.[3] Because of this fluidity over time and space, health is not a fixed state of being but a dynamic and unfolding process. Health is an expansive ideal that does not attend simply to the presence or absence of disease; rather, it is the achievement of well-being and the capacity to live well. This includes the attainment of material goods needed for a quality of life but also physical comfort and positive mental health. Health is achieved when people have access to the means of making a healthy life, and it is differentially experienced because of disparities within societies that result in unequal vulnerabilities and deprivation.

Lastly, while both of the stories highlight the social conditions underpinning health and well-being, less is said about either the environmental or the ecological dynamics that contribute to the spread of infectious disease and the possibilities for human health. Absent from these accounts are the environmental conditions in American Indian reservations that undoubtedly play a role in producing the social and economic opportunities for residents to generate income and meet individual needs. Local views on the legacy of land dispossession are briefly mentioned but not examined as a central factor shaping contemporary circumstances. Similarly, the South African article emphasizes the role of the national government in either responding to or failing to address the HIV/AIDS epidemic; however, no attention is directed to the historical environmental conditions that contribute to the spread of infectious disease or that shape the possibilities for those infected and affected by HIV. Rather, South Africa's complicated and deeply contested spatial history is overlooked to allow the far easier assignment of blame for the epidemic to a governmental administration or, more tellingly, one of its representatives.

As evidence of the dynamic nature of human health, the management of a disease such as HIV has changed in many parts of the Global South because of an aggressive push to universalize access to ARVs. This expansion of access to these life-saving drugs has made it more common for academic and policy communities to refer to HIV as a chronic condition, akin to diabetes or heart disease. But chronic HIV, or what I refer to as "managed HIV," entails other challenges. Through many research visits to South Africa, informants have shared with me, time and again, that while they are less likely to immediately die from AIDS, the strain on their mental health is considerable. Often it is women, alone in managing their health and the well-being of their family, who discuss the pressures they face in meeting basic needs, including securing food to maintain the effectiveness of ARVs. In many of these conversations, HIV recedes to the background while other concerns become paramount. Access to life-saving drugs, while a miraculous development from a previous time when the virus meant certain death, does not erase all of the challenges for those *living with HIV.*

Space profoundly shapes the possibilities for human health. My central objective is to demonstrate how spatial processes result in dynamic health domains for individuals, communities, and regions. The production of space involves the interplay between people and disease patterns that converge in distinct places and landscapes. This includes the built environment, historical processes that situate people in particular locations, and symbolic and cultural meanings. Yet, while there has been extensive research on the social determinants of health, ecological conditions are also integral to the spread of disease and health management. Whether it is exposure to toxins in the industrialized North or the spread of infectious disease in the Global South, existing scholarship locates pathogens, vectors, individuals, and populations within ecological systems that shape the places and landscapes of health. It might be tempting to emphasize either social or ecological systems as producing the states of disease; however, it is these social ecologies that create the possibilities for human health. I refer to these points of intersection as the *political environmental context.* The political environmental context is produced through spatial processes that generate differential vulnerability to the spread of infectious disease and exposure to noninfectious disease. Rather than address a single disease pattern, I engage with cases of infectious and noninfectious disease to show how space shapes the ways in which health is embodied, experienced, and managed.

This book was conceived and written over a period of years that have seen an expansion of interest and an opening of theoretical engagements in the social and natural sciences on the subject of human health. Spanning a diversity of disciplines, including geography, anthropology, public health, and epidemiology, research and policy attention has been directed toward understanding the role of social processes in shaping the conditions under which diseases spread and in which health is understood and best managed. As a leading example of this, the physician and medical anthropologist Dr. Paul Farmer has shown how structural conditions, whether they are political, economic, or cultural, make certain social groups more vulnerable to infectious disease or illness. Farmer asserts that health inequities should be interpreted as a form of structural violence that constitutes a social injustice.[4] As a result, it is increasingly common to locate social actors in broader social, political, economic, and cultural systems to explain the ways in which disease vulnerabilities are created and how the possibilities for health management are enabled or constrained. Invaluable as this has been to studies of human disease, the role of spatial processes is often unaddressed, as are the ways in which social and ecological systems interact in shaping human health. This is why I emphasize the political environmental context as central to health and well-being, specifically in producing differential vulnerability to the spread of infectious disease and exposure to the conditions that generate noninfectious disease.

The role of social and ecological systems in generating health outcomes has been a subject of interest in a number of fields. Scholarship in the social sciences has revealed the links between humans and the natural world, presenting multiple knowledges and the agency of nonhuman species. The production of economic relationships, facilitated through the global development process, has resulted in particular outcomes for human health, whether it is the improvement of access to certain services and medications or an increasing vulnerability to the spread of disease due to large infrastructure projects such as the construction of a hydroelectric dam. The concept of the "epidemiological transition" asserts that socioeconomic development results in social and cultural transformation that reduces the risk of infectious disease, though it has been shown that this can be followed by an attendant increase in chronic illnesses. The epidemiological transition has also been challenged for neglecting to consider the double burden of illness in the Global South, whereby infectious and noncommunicable diseases are present.[5] In addressing the ecological determinants of health, ecology and biology have emphasized a range of biotic and abiotic factors that facilitate the spread of infectious disease or the production of conditions that influence noninfectious disease. This is contributing to varied understandings of human health; however, there remains a need to more fully integrate social and ecological perspectives in order to recognize the diverse factors shaping differential health possibilities.

While engaged with several of these intellectual fields, I am also deeply committed to fieldwork in South Africa, where I have worked since 1999. My orientation has been shaped not by a single theoretical perspective but by the desire to adequately describe what I have observed through research that has spanned more than a decade. Having now lived and worked in a country that has experienced tumultuous periods over its own states of disease, I find myself concentrating on the conditions that have produced, and continue to produce, human health. These conditions are often the product of spatial legacies that result in differential vulnerabilities to disease and exposure to the factors that undermine health and well-being. I begin *States of Disease* by detailing some of the features of the HIV/AIDS epidemic in South Africa during an especially contentious period, but I later return to more recent work in the country to describe the ways in which the lived experiences of the epidemic have changed, either in terms of social perceptions of HIV or in the ways that the state and citizenry confront each other in its wake.

The HIV/AIDS epidemic has now become endemic in the country, largely through improvements in treatment that have made HIV lifeways for many akin to a chronic condition. These encounters have taken on new forms as HIV has moved from certain death to managed state. Yet while recent popular reports such as a cover story in the *Economist* question whether the global community has approached the "end of AIDS," HIV remains the lived experience for millions and continues to transform the social and ecological systems for those infected and affected.[6] Simply put, the improvements in managing HIV do not erase the social ecologies in which the virus survives and spreads. While fewer people die from AIDS in South Africa than a decade ago, they remain differentially affected as a result of social dynamics such as familial support networks, gender, and capacity to access governmental assistance. The landscape of HIV has been transformed by greater access to ARVs; however, it remains an uneven terrain that people are differentially equipped to navigate. Central to this book is the desire to address the collision of factors that makes human health dynamic for individual bodies, families, communities, and populations. While I begin with a discussion of HIV/AIDS in South Africa, this book is not solely about this epidemic. Rather, I consider the ways in which human health is produced, understood, and managed in various settings. Although concerned with the lived experiences of disease, I shift the focus away from individuals and isolated illnesses and toward the spatial processes that make people differentially vulnerable to poor health. In doing so, I challenge conventional understandings of disease in order to envision human health in more holistic, equitable, and just ways.

It will be clear throughout this book that my training is situated in the social sciences, with a particular emphasis on geography, development studies, and social and environmental justice. While my work takes seriously the ways in which biophysical processes contribute to the possibilities for human health and well-being, I concentrate here on the intersections between social and ecological processes, specifically within the political environmental context. While my argument is theoretical, it is deliberately empirical so as to be guided by my informants. I do not presume to speak for them, but rather engage with their states of disease and the social ecologies producing them. Human health is complex. Its definition depends on the object of study, and it changes over time because of the structures that people inhabit. These are social structures to be sure, but they are also ecological, and through their intersections the political environmental context helps determine which people are more vulnerable to illness, able to access services and amenities that contribute to human health, and, in the case of severe epidemics, able to access medications for survival. Human health is a complex and contextual process that is shaped by many factors, but central to our understanding should be three simple questions: Who gets sick? Why are they sick? And once they are sick, does their health improve? This book addresses these questions with the belief that only by doing so is it possible to reduce inequities in disease exposure and expand the opportunities for human health and well-being.

Acknowledgments

This book is the result of innumerable conversations and exchanges, in addition to generous readings and rereadings by friends and colleagues. It began in December 2009, though I must admit I didn't know it at the time. I shared material that would eventually become the introduction and chapter 1 with Brent McCusker, Ed Carr, Jeff Bury, and Molly Brown, all of whom provided critical feedback for which I am grateful. Lorraine Dowler, Melissa Wright, and James McCarthy provided invaluable insights on several of the chapters. While still in its early stages, the book benefited from the input of Hannah Love at UC Press. Hannah recruited this manuscript, and I am grateful for her early encouragement. As the writing continued, Naomi Schneider was my primary editor and was supportive at all stages. I am thankful for her many helpful inputs and feedback as the work was completed. My thanks to Elaine Guidero, who assisted with the development of figure 2; and to Aaron Dennis, who completed figures 1, 3, 4, and 8.

I benefited further from interactions with various people and academic groups at the Pennsylvania State University (Penn State), where I presented earlier versions of this material. I also received feedback from individuals at West Virginia University and the Center on Health, Risk and

Society at American University. As the manuscript took on fuller form, it was improved by the close readings of several people, including Meg Winchester, Ben Marsh, Erica King, Richard Earles, Nari Senanayake, Abigail Neely, Joel Wainwright, Brent McCusker, and an anonymous reviewer. Further development of my thinking about health–environment interactions resulted from coediting *Ecologies and Politics of Health* with Kelley A. Crews. I appreciate our conversations and exchanges of ideas on that volume, and I also thank the contributors for their work in pushing my thinking on the ecological dimensions of health and vulnerability, the sociopolitical dimensions of human health, and the intersections between the ecological and social dimensions of health. I also benefited from a graduate course at Penn State titled "Political Ecology and Human Health." My thanks to the students who participated in the seminar, and to the many others with whom I work who have contributed to my engagements with human disease and health.

Funding to support the research and writing of this book was generously provided by four separate research grants from the National Science Foundation (NSF). The majority of research reported in this book was supported by an NSF CAREER grant (BCS/GSS no. 1056683). Two undergraduate research assistants from Penn State, Evan Gover and Marina Burka, were funded through Research Experiences for Undergraduates supplement awards that allowed them to conduct fieldwork in South Africa in 2013 and 2014, respectively. I am also indebted to the Department of Geography and the College of Earth and Mineral Sciences at Penn State for providing funding at various stages of research in Botswana and South Africa. I am fortunate to be in an environment that encourages this type of work while valuing the endeavor of book publishing. The completion of the manuscript was supported by a fellowship from the Social Science Research Institute at Penn State, and by an academic sabbatical during which I am currently affiliated with the Department of Environmental and Geographical Science, and the African Climate and Development Initiative, at the University of Cape Town. It is immensely gratifying to finish this book in the country in which it began.

The research conducted in Botswana in 2010 was supported by a RAPID grant from the NSF (BCS/GSS RAPID no. 0942211). This led to a three-year research project also supported by the NSF (BCS/GSS no.

0964596). The latter project is examining how perceptions of environmental variability prompt livelihood adjustments, particularly diversification into new forms of agriculture and resource collection that rework access patterns for residents. Although I conducted all the interviews reported in chapter 5, I have benefited from engagements with others, particularly Kelley A. Crews, Kenneth R. Young, Thoralf Meyer, Jamie Shinn, and Kayla Yurco. I am also indebted to Gully and Kenny, who assisted with the translation of the interviews in the Boteti region.

While this book doesn't draw directly on material published elsewhere, I have written about South Africa's historical systems of spatial segregation, with particular attention to the Bantustan system, in other venues. As such, some of the material in chapter 3 is informed by research I completed for articles in *Transactions of the Institute of British Geographers*, *Area*, and the *Geographical Journal*. My article "Political Ecologies of Health" that was published in *Progress in Human Geography* provided an early framework for some of the arguments that I make in this book, although fieldwork and continued writing have extended my thinking. In particular, I appreciate the contributions of political ecology to subjects of human disease and health, but I draw from other disciplines in advancing the "social ecology of health" framework articulated herein.

Three of the chapters rely on fieldwork completed in South Africa at intervals from November 2012 to June 2016. My postdoctoral scholar in health and environment interactions, Dr. Margaret (Meg) Winchester, assisted with the recruitment of the focus-group participants and the interview guide for the sessions conducted in 2012. Meg also oversaw the completion of the focus-group interviews, and I am grateful for all of her assistance. Two members of my research team, Cliff Shikwambane and Wendy Ngubane, facilitated these interviews. I have had the great fortune of working with Cliff and Wendy since 2002, and I am indebted to them for their hard work and commitment to multiple research projects. Erens Ngubane has also remained a close friend and invaluable collaborator while helping guide the team through much of the research discussed in the book. The structured surveys were conducted in June–July 2013 by Wendy and Erens, and also by Tsakani Nsimbini, Golden Nobela, Ncobile Thumbathi, and Cindy Nhone. Meg was instrumental in the training of the research team and overseeing the completion of the survey. She was ably assisted in the field by research assistant Evan Gover. Sadly, Cliff passed away in 2013 following complications from a car accident. This was a heartfelt loss for me and the rest of the research team. He is deeply missed and his family is in our thoughts and prayers.

Throughout the book I refer to interviews with residents of communities in Botswana and South Africa. In order to protect their identities, I use pseudonyms even in cases where they gave their permission to be identified. That said, even though I choose not to name them in the book, I am deeply thankful to the many people who took time to speak with me and other members of my research teams over fifteen years. Their generosity is appreciated beyond measure.

Finally, I must thank the love and support that I have consistently received from my wonderful family. My parents, Tom and Kathy Wetherell and David and Marsha King, have always encouraged my pursuits. Randy and Marty Locke have been supportive in so many ways. My children, Madigan and Aidan, have been a source of joy and inspire me with their optimism about the future. I have laughed much more during the writing of this book because of them. Most importantly, my wife, Erica King, has been unfailingly supportive throughout this undertaking. Thank you for making this possible.

Abbreviations

AIDS   Acquired immunodeficiency syndrome

ANC   African National Congress

ART   Antiretroviral therapy

ARVs   Antiretroviral drugs

HIV   Human immunodeficiency virus

IPCC   Intergovernmental Panel on Climate Change

NGO   Nongovernmental organization

RDP   Reconstruction and Development Programme

WHO   World Health Organization

Introduction

News reports and images from Africa and other world regions have made various infectious diseases a global concern. The appearance of a drug-resistant strain of tuberculosis in South Africa in September 2006 raised fears about the health effects for those living in poverty, in addition to the potential complications for the millions in the country already suffering from HIV/AIDS. A different health crisis began unfolding in Zimbabwe in 2008, as thousands of people were infected with cholera because of the breakdown in infrastructure and sanitation systems that accompanied the country's economic collapse. Threats from H1N1 (swine flu) and avian flu strains have made the word *pandemic* commonplace while offering reminders of the interconnectedness and vulnerability of the global community. Popularly written books raise alarms about the transmission, or spillover, of infectious animal diseases to human populations.[1] Widespread media coverage beginning in August 2014 detailed that Ebola had crossed the shores of the Atlantic, carried by two Americans who were part of a humanitarian group working to address the outbreak in West Africa. Their health improved quickly after receiving an experimental treatment in Liberia that had previously been used only on animal subjects. Additional care was administered in a specially built isolation unit in

1

Emory University Hospital in Atlanta, Georgia. Several African governmental officials and public health experts responded to this development by decrying the differential treatment, asserting that the thousands infected within West Africa should be given the same level of care.[2] Days later it was announced that the experimental drug Zmapp would be provided by Mapp Biopharmaceutical to Liberia at no cost, though the company indicated that this provision had exhausted the supply of the drug.[3] After several months, anxiety within the United States dissipated even while the outbreak alarmingly continued in West Africa into 2015, resulting in the deaths of more than eleven thousand people.[4]

Global health is also implicated in the political and economic dynamics associated with international development and geopolitical alignments. Multilateral institutions, including the United Nations and the World Health Organization, have insisted that more funding is needed to stem the increasing number of deaths from HIV/AIDS and other infectious diseases. The United Nations Millennium Development Goals committed resources to preventing diseases such as HIV/AIDS and malaria, and while there have been marked improvements, the ambitious goals were not fully achieved by 2015. This is all the more troubling given the numerous ways that disease disrupts social and ecological systems. Human diseases remove family members and strain social relationships that are critical to community survival, while destabilizing economies, ecosystems, and agricultural productivity.[5] The loss of a family member eliminates parents from caregiving duties, thus transferring the responsibility to young adults or elderly family members. Household incomes are reduced, which places greater pressure on children to seek employment rather than attend school. This has long-term effects for families and societies that continue for decades after the death of a family member. Disease also reshapes the relationships between human populations and the natural environment, while threatening the possibility for sustainable decision making. The availability of natural resources often declines as a result of illness, and land-use strategies shift to generate income or to meet household needs in the short term.[6] The spread of human diseases, therefore, has tremendous impacts on social and ecological systems and their coupled interactions. But it is also through these social and ecological interactions that the possibilities for human health are created.

I have observed these patterns over fifteen years while living and working in South Africa. In August 2001, I began my doctoral research in
Mpumalanga Province in the northeast of the country, not far from the
borders of Swaziland, Mozambique, and the Kruger National Park, which
is a major tourist destination for international visitors. I had first visited
South Africa in 1999 to conduct fieldwork for my master's thesis in
geography. This was less than five years after the democratic elections,
and a palpable energy and optimism infused the country. At that time, I
was intent on examining how civil-society organizations were reorganiz-
ing themselves as part of the transition, given that many were previously
working in opposition to the apartheid government. I returned in 2000
for six weeks of preliminary research, with a particular interest in the legacies of colonial and apartheid spatial regulation of social and environmental systems. The intention of my doctoral research, which took place
over one year (2001–02), was to examine how individuals, households,
and communities in rural areas were experiencing transformations
following the 1994 elections. After more than four decades of apartheid
rule, the first fully representative elections in South Africa's history
brought President Nelson Mandela and the African National Congress
(ANC) into power. This was a momentous event, with the entire world
watching as the relatively peaceful elections swept aside the apartheid system of racial segregation. Overseas commentators and political observers
marveled at the long lines on the day of national elections that stretched
out by the hundreds, ensuring for many that it would take hours to cast
their first ballot. President Mandela and the ANC presented the new
South Africa as a Rainbow Nation that embraced the divergent populations in the country, which was epitomized by the new flag that displayed
the colors of green, blue, black, white, gold, and red in an orchestrated
combination of movement from left to right. This new South Africa was
presented as a place of hope, optimism, and opportunity for the millions
who had been disenfranchised by centuries of colonial and apartheid rule
that utilized racial classification and segregation to manage the majority
of the country's population for the benefit of the white minority.

The political transition that accompanied this period was marked by
tremendous challenges to overcome centuries of racial segregation, which
reached its nadir with apartheid "separate development" policies in the

1960s and '70s. Apartheid was a social, political, and economic system, but it was made possible through the production of space. Building upon the colonial establishment and administration of native reserves, the apartheid state exercised space as a mechanism of social control. Central to this was the construction of ten Bantustans, or homelands, which were territories that the minority white state had demarcated for the majority African population. Rural areas became the site of subtle and overt violence, as national and governmental authorities utilized various strategies to segregate and regulate the population. The Bantustans, taken together with the townships that ringed the urban centers of the country, constituted an elaborate geography of racial segregation. Although it is impossible to be certain, some studies estimated that during apartheid more than three million people were forcibly moved into the townships and Bantustans, and strict controls were instituted to regulate migration and economic activity.[7]

My research at that time was focused on understanding how social, economic, environmental, and spatial changes were affecting livelihood systems, the governance of natural resources, and the ways in which economic development was being conceptualized by provincial and national agencies in the post-apartheid era. With such a deliberately engineered social organization, I was fascinated to understand the legacies of these spaces—and possibilities for political and economic transformation. I concentrated my research in Mpumalanga Province because part of the region had been territory within the KaNgwane Bantustan during the apartheid era and therefore served as a microcosm for other areas in the country. KaNgwane had been established in the 1970s through the creation of the Swazi Territorial Authority, which had jurisdiction over the region. This was part of the apartheid ideology of "separate development," whereby the ten Bantustans were presented by national authorities as existing along a development trajectory, essentially a linear pathway that would conclude with their independence from the South African state. Separate development was central to the discourse used to justify racial segregation by arguing that the Bantustans were being liberated to pursue their own path.

In practice, however, the material features of separate development involved a paternalistic developmentalism that reified active interventions by state authorities.[8] National propaganda documents asserted that separate development was justified because the majority African population had the right to independent and self-directed development, and that it was the responsibility of the white state to facilitate this process. As one example, the government report titled *Black Homelands in South Africa* explained that the apartheid government was moving toward a "solution which accords each group its inalienable right to determine its own destiny and formulate its own scale of values."[9] In a similar rhetorical sweep, the South Africa Information Service explained that "each of the different black nations in South Africa, we believe, should have the opportunity to exercise its basic right to determine for itself its own future. Nothing should prevent each of these black nations from becoming independent in the fullest sense."[10] These grandiose statements ran counter to the political and economic contradictions of apartheid spatial development, which by the late 1980s was increasingly recognized by members of the national government as unsustainable. This prompted engagements with the ANC leadership to work toward a peaceful transition, and the release of Nelson Mandela from Robben Island prison in 1990 signaled that apartheid's days were numbered.

The dismantling of the apartheid system in anticipation of the democratic elections resulted in the incorporation of the Bantustan territories and their populations into the national geography. A central element of my research was to examine the effectiveness of a community-based ecotourism project in generating employment and other benefits to the community. The Mahushe Shongwe Game Reserve was managed jointly between the Mzinti community and the provincial conservation agency, the Mpumalanga Parks Board, and served as an entry point into processes of livelihood production, governance, and development in the new South Africa. As part of my work, I spent eleven months in the Mzinti community. I observed meetings and markets, talked with local leaders of the tribal authority, and used aerial photography and community-mapping exercises to understand how the community was constituted and whether it was experiencing change in the wake of the democratic elections. Interviews were conducted with household members about their family history and livelihood systems.[11] Although my research was not initially centered on infectious disease, it became painfully clear that HIV/AIDS was transforming rural communities such as Mzinti. HIV/AIDS was a *disease hidden in plain sight*. People were dying from AIDS, but there was little public discussion about the disease. Rather, deaths were being attributed to a variety of illnesses associated with late-stage AIDS, most notably tuberculosis. Visits to funeral homes and local clinics revealed that deaths were increasing, but few people would admit that AIDS was the cause. Working with several community members, my research team and I conducted a structured survey of 478 households in the Mzinti community. The survey included detailed questions on household demographic patterns, dependence on natural resources, and mortality rates. In one of the more transparent signs of the disease's invisibility, among households that had reported the death of a family member during the previous five years, only one household attributed the death to AIDS. Yet dozens of households reported a death from tuberculosis. Other respondents indicated that they were visiting *sangomas,* who are one type of traditional healer, for various treatments, including the use of the African potato that was being prescribed for AIDS-related symptoms. Residents were uncomfortable speaking of the disease openly, and there were concerns that stigmas were being attached to HIV-infected individuals, particularly women, throughout the country.

The impacts of HIV/AIDS were also observable through cultural practices, some of which appeared to be eroding because of the increasing number of deaths. One example of this was livestock ownership and its relationship to funerals. Even though only a fraction of the community owns cattle or goats, the possession of livestock is important both economically and culturally to community members.[12] Livestock are "stored capital," much like a bank account, that can be drawn upon in the case of emergency or to meet the needs of the household. Interviews with residents revealed that cattle would be sold to pay for school fees or uniforms, or for income in cases where a family member had lost employment. In speaking with livestock owners, I learned that the slaughter of a cow would typically accompany the funeral of a family member. A traditional funeral occurs over the duration of a weekend, and it is common for family members and friends from outside the community to travel and stay for the service. A *braii* (barbecue) takes place, and meat with corn porridge is served to the guests. In talking with community members, it was clear that the practice of slaughtering a cow to accompany funerals was coming into conflict with the increasing number of deaths. As one livestock owner explained to me: "Cattle are very expensive. When you want to buy a cow, it is very, very expensive. For example, when there is a death in the family, there must be a cow, which is slaughtered. It does not matter whether it is a child or adult." She went on to explain that cultural practices were undergoing a change because of the increasing death rates and continued poverty in these areas, noting:

> [Cattle ownership] is part of the tradition, but sometimes it looks like it is going to go away because some of the people are very poor. They don't have money to slaughter a cow, then you find that some other people will not go to attend that funeral because they won't eat. When someone dies you don't have enough money because you have to buy coffins and prepare for the funeral. Therefore, there's no need to prepare food for the people. I attended a funeral last week in Barberton. We didn't have anything, we only just went to the graveyard, then we came and with our hands we prepared [the food]. They didn't have anything so we had to make a contribution for the coffin.

In addition to livestock, other livelihood practices were coming into direct conflict with the rising number of deaths in the community. Because of the high population densities in the former Bantustan territories, agriculture was not viable for the majority of residents. The collection and use of natural resources, however, was more common, and something that individuals and households relied on to meet subsistence or income needs. After completing the survey of Mzinti households, it was clear that the collection of wood from communal areas surrounding the community was the most common resource dependency. More than half of the households surveyed reported cooking with wood on a regular basis, and in many cases it was their primary source of energy. But other resources, including thatch grass, medicinal plants, and sand, would be collected at various times for different purposes. When I began my doctoral fieldwork in August 2001, residents of the community were allowed to enter the Mahushe Shongwe Game Reserve to collect wood to be used for a funeral. In order to gain access to the reserve, which was fenced around its perimeter and closely monitored by the Mpumalanga Parks Board (later renamed the Mpumalanga Tourism and Parks Agency), residents would obtain permission from the local tribal authority official. Reserve managers reported that wood collection was on the rise in Mahushe Shongwe, and they did not think that extraction was sustainable. In returning to the community in 2004, I learned that wood collection was no longer allowed within the reserve. Reserve managers explained that the number of funerals in the community had made it necessary to end this practice, stressing that otherwise the demand would invariably degrade Mahushe Shongwe. The likely consequence is that the resource pressure moved to the communal areas, as people continued to collect wood, medicinal plants, and other types of resources to support household livelihoods disrupted by the sickness or death of a family member.[13]

The seeming contradictions within local communities were reflected in the national sphere as well. Increasing concern about the national response to HIV/AIDS was reaching full boil, with various civil-society organizations pressuring the South African government to provide access to antiretroviral drugs (ARVs). Most notable was what critics were calling a "denialist" position within the ruling ANC government that was being reflected by public statements and the intransigence in distributing ARVs.[14] The denialist position was marked by statements made by some governmental officials that questioned either the direct links between HIV and AIDS or the benefits of particular treatments, including nevirapine, which had been shown to prevent mother-to-child transmission of HIV. Leading members of the ANC were suggesting that nevirapine was toxic and that further tests were needed to determine its efficacy before widely administering it within the country. By 2002, the Treatment Action Campaign (TAC), a non-governmental organization (NGO), had been waging a public relations campaign to force the rollout of ARVs to the population. Specifically, TAC sued the government by asserting that the denial of access to nevirapine was a violation of the Constitution.[15] News reports documented the arrival of a boat from Brazil that had landed outside Cape Town and was distributing drugs in defiance of the mandates of the South African state.

Meanwhile, key governmental leaders, including the former president, Thabo Mbeki, received public condemnation when Mbeki asserted that HIV should be seen as part of a gamut of diseases that disproportionately affected the poor. The former health minister, Dr. Manto Tshabalala-Msimang, regularly emphasized that HIV/AIDS was not just a health problem, but a development challenge as well. Dr. Tshabalala-Msimang also stated that traditional medicines, including the African potato, garlic, and beetroot should be part of treatment at certain stages, eventually garnering her the nickname "Dr. Beetroot" from members of the global health community. In describing the state response to HIV/AIDS at this time, South African scholar Peris Jones argued that AIDS was presented by governmental actors "as a syndrome rather than a disease, implying that people do not actually die of AIDS but [of] poverty and the opportunistic infections it nurtures."[16] The need to distribute ARVs was also viewed by some as a neocolonial fiction constructed by Western donors and corporations for the purpose of selling pharmaceuticals.[17] These views were understandably intensified by the intransigence of many of these companies to provide generic ARVs, which would have required breaking international patent protections. It is important to be clear on this point. As I discuss throughout *States of Disease*, the linking of human disease with socioeconomic poverty is necessary for understanding the production of health vulnerabilities. People are more susceptible to infectious disease if their nutritional and caloric needs are not met. Additionally, socioeconomic poverty in the United States is often directly linked to exposure to environmental toxins that contribute to chronic health conditions such as asthma, bronchitis, or pulmonary edema. And yet, set within the context at the time, some critics accused the South African government of coupling poverty with disease as a strategy to delay the distribution of ARVs.

In 2004 and 2006, I spent more time engaging with governmental officials and nonstate actors about HIV/AIDS. After years of researching a range of social and environmental topics, it was clear that for those infected or affected, the epidemic was touching nearly every facet of life. Understandings of the disease had become more public in the ensuing years, even as the national government continued to resist the sort of aggressive response that other African governments had pursued.[18] Several officials, including Dr. Tshabalala-Msimang, had become pariahs on the international stage, and at the end of the international AIDS meeting in 2006, Stephen Lewis, the United Nations Ambassador to Africa for AIDS, pronounced that South Africa "is the only country in Africa whose government continues to propound theories more worthy of a lunatic fringe than of a concerned and compassionate state."[19] Meanwhile, the rates of estimated HIV infection had increased and there existed clear regional disparities, some of which appeared connected to historical spatial systems, including the Bantustans. Research shows that outmigration from rural areas to seek employment in the mining sector contributed to increases in the transmission of the disease,[20] and that social inequalities in income and employment status resulted in greater exposure to risky sexual activity, diminished access to health information, and delayed diagnosis or treatment.[21] While the country as a whole has a reported 18 percent HIV infection rate, areas within Mpumalanga Province where I had been working were cited by some provincial health ministers as having a 45 percent estimated infection rate. These estimates were echoed by other civil-society actors working in the region to emphasize the particular geographies of the epidemic.[22] Much of the long-term care for HIV-infected populations at that time was being administered by home-based care workers, many of whom were volunteers, and local officials complained that the United States President's Emergency Plan for AIDS Relief (PEPFAR) was placing restrictions, such as a requirement for family-planning education, upon the dispersal of resources. In some cases, NGOs that received funding had to specify that they would advocate family planning to patients.[23] Other nonstate groups continued to protest the role of pharmaceutical companies in suing to protect patents that restricted the distribution of generic drugs to sick people.

After years of intransigence and denial about the severity of the HIV/AIDS epidemic, South Africa now has a more hopeful future. The 2008 World AIDS day was particularly notable in coinciding with a significant political change in South Africa's national response to HIV/AIDS. The resignation of President Mbeki in September of that year, and the subsequent replacement of Dr. Tshabalala-Msimang, heralded to many a new beginning as the government moved away from years of denial and resistance to the distribution of ARVs. In a sharp break from previous governmental policies, then Health Minister Barbara Hogan pledged to reduce by half the number of new infections by 2011 and ensure that 80 percent of the people with HIV/AIDS received treatment and care. Coinciding with the 2009 World AIDS Day, President Jacob Zuma promised more aggressive governmental responses to the disease. One year later, he tested publicly for HIV and disclosed his status in an effort to encourage more open discussion about the disease.[24] Sadly, these shifts came shortly after a report from researchers at the Harvard School of Public Health that estimated that governmental delays in distributing AIDS drugs between 2000 and 2005 resulted in the deaths of at least 330,000 people.[25] Additionally, the study reported that 35,000 babies were born with HIV during the same period because of the government's reluctance to introduce a mother-to-child transmission prophylaxis program using nevirapine. The authors of the report were explicit in stating that the South African government acted as "a major obstacle in the provision of medication to patients with AIDS."[26]

Clearly, there are many points of entry to understanding a disease like HIV/AIDS. But one particular moment stays with me as I reflect on my years in South Africa. My time in rural communities led to many anguished conversations with colleagues to try to understand the situation. One particular discussion was with a German development official assigned to the Office of the Premier for Mpumalanga Province. I remember describing in detail what I was seeing in the rural areas, and how difficult it was to reconcile these findings with the national debates and failures to aggressively respond to the disease. After listening to me for several minutes, the official paused and said very simply, *"In South Africa, no one dies of AIDS."* That comment has continued to haunt me, largely because it seemed a tragic, albeit fitting, way to describe the states of disease in South Africa. For how else to explain a disease that was affecting millions of lives yet remained elusive in public debates and largely hidden from plain view? How else to understand a disease whose interpretation through social and cultural practices and discourses seemingly challenged the hegemony of Western biomedicine or was connected to critiques of development policies that contribute to the production of socioeconomic poverty? How is it that an infectious disease that is labeled by some as the product of HIV transmission from one individual to another has come to be labeled by others as the product of a neocolonial Western medical complex that financially benefits from the sale of pharmaceuticals? What do these conflicting views and normative positions mean for how we understand and respond to infectious disease? What do they mean for how we understand human health?

HEALTH ECOLOGIES WITHIN DYNAMIC SYSTEMS

Expanding our scope to Sub-Saharan Africa reveals other health challenges for human populations and the social and ecological systems in which they are embedded. While future global climate change is impossible to predict with certainty, most scientific models suggest that increasing variability will accompany rising temperatures. For example, the Fourth Assessment Report of the Intergovernmental Panel on Climate Change (IPCC) indicated that southern Africa is especially vulnerable to increases in environmental variability, particularly in terms of precipitation and flooding.[27] Rainfall may come sooner or later in the season, and the amount that arrives at any given time may be more variable. The consequences for agricultural production could be severe, challenging nutritional health in families already near or at the state of poverty. Disease vectors—the organisms that transmit particular infections from one entity to another—will shift as a result of these conditions, in ways that will have unanticipated costs for human populations. Increased vulnerabilities to infectious disease and opportunistic infections are anticipated to result, thereby demonstrating the closely linked relationships between humans and the natural environment, relationships that are essential to human and ecosystem health. Recent reports from the biological sciences demonstrate that small changes in temperature, even on a daily basis, can affect parasite infection, the rate of parasite development, and components of mosquito biology that combine to determine the intensity of malaria transmission.[28] The same has been shown for Dengue virus transmission by the mosquito *Aedes aegypti:* its life span and susceptibility to Dengue infection decrease when there are larger fluctuations in daily temperature, as compared with only moderate fluctuations.[29] The implication is that human and nonhuman species are directly affected by changes in the natural environment, and their coupled relationships will likely shift in the future.

Changing environmental conditions, or more dramatic variabilities in ecosystem functioning, have material as well as symbolic importance for the ways in which people perceive and manage their health. Since 2007, I have been conducting research in the Okavango Delta of northern Botswana, a setting that experiences often significant variability in precipitation and flooding patterns. Beginning in 2008, increases in precipitation pushed water to reach tributaries to the south of the delta, including the long-dry Boteti River, which resulted in some residents experiencing water for the first time in twenty years. Working with colleagues from the University of Texas, with support from others in Botswana, we conducted interviews with people who lived along the river. These residents, who often pumped water from beneath the dry riverbed using gasoline-powered generators, were interviewed at various points along the river, and while some had yet to see the water, others had just experienced it for the first time the previous year. In lengthy conversations, they spoke about the return of the water and the ways in which it was disrupting and reshaping the lived experiences of families. Cattle *kraals* that had been established in the riverbed were being relocated. Dryland farming and floodplain farming, known as *molapo*, were also adjusting in response to the dynamic ecological conditions, either to reduce the negative consequences or to take advantage of the opportunities presented by increased proximity to water. The impacts for household economies and livelihood systems were anticipated to be significant and would unfold over time with the rising Boteti.

An unanticipated finding was the frequency with which people coupled views of the water's return with their understandings of health—either their own or that of their livestock. Lumpy skin disease, which can be fatal to cattle, was raised as a concern, and many speculated that it would increase in severity because of the water's return. Local residents also talked about threats related to swimming in the river or from conflicts with wildlife such as crocodiles and hippos, or the possibility that malaria would become more prevalent. These interviews revealed in detail how local perceptions of health, and understandings of potential risks and vulnerabilities due to climatic and ecological variability, were integrated with the return of the Boteti River. As discussed further in chapter 5, these findings raise several questions in regard to human health, including the following: How do human–environment interactions shape human health? How do environmental changes affect human health and the perceptions that shape health decision making? What do future climate change and environmental variability mean for human exposure to disease vectors? In addressing these questions, I will examine what shifting health ecologies mean for disease, livelihood systems, and environmental health.

States of Disease addresses the questions raised by these cases to engage with and challenge received wisdoms about human health. In drawing upon the seemingly disconnected cases of HIV/AIDS in South Africa and environmental change in northern Botswana, I concentrate on how human health is shaped by spatial legacies that generate inequities in the places and landscapes of health. The inclusion of these cases is designed to advance my central argument that human health is the result of the political environmental context that produces inequities in exposure to infectious disease and the conditions that contribute to noninfectious disease. The political environmental context results in unequal access to health-care services and the ability of health-care agencies to effectively respond. It is produced by the relationships between social and ecological systems that converge in generating both disease and the opportunities for human health and well-being. While this contention may not surprise those who study the social determinants of human health, there remains a tendency in research and policy communities to minimize the role of sociopolitical factors. Rather, disease is often presented as a "natural" event that occurs because of ecological conditions or individual decision making. This tendency is derived, in part, from the Western biomedical tradition, which presents disease as a result of exposure to microbes while presenting humans primarily in biological terms. While research in the social sciences has attended to the social dimensions of human health, there have been varied understandings of the roles of political economy, power, and discourse in shaping health and well-being. Additionally, my approach emphasizes the role of ecological processes in the spread of infectious disease or in producing health vulnerabilities, which can similarly be underemphasized by studies on the social determinants of health. *States of Disease* advances a "social ecology of health" framework to argue that human health is shaped through the intersections between social and ecological processes that structure the spread of infectious disease and exposure to noninfectious disease.

STATES OF DISEASE

It is important to emphasize at the outset that my focus is not on one particular disease pattern but on human health in its fullest sense. This resonates with a number of developments over the past several decades. For example, the Constitution of the World Health Organization (1946) defines health not as the absence of disease or infirmity but as "a state of complete physical, mental and social well-being."[30] It identifies health as a human right, irrespective of social, economic, religious, and ethnic differences and states that "the health of all peoples is fundamental to the attainment of peace and security."[31] In another notable advancement, Canada's Lalonde Report (1974) proposed the concept of the "health field" to address the causes of sickness and death, in order to facilitate public health interventions and improve human health. The report identified four elements, specifically human biology, environment, lifestyle, and health-care organization, as being equally important in ensuring a comprehensive health policy.[32] Lastly, the Ottawa Charter for Health Promotion (1986) discussed health as including the realization of aspirations, the satisfaction of needs, and the ability to adjust to the environment.[33] The Ottawa Charter specifically identified the following as fundamental conditions for human health: peace, shelter, education, food, income, ecosystem stability, sustainable resources, social justice, and equity. I draw upon these traditions in understanding human health not as the absence of disease but as the achievement of well-being and the capacity to live well, which includes the attainment of suitable and sufficient material goods. This perspective is also concerned with the roles of external stress and mental health, while emphasizing that individuals have the agency to make decisions and contribute to their own lifeways. It attends to how disease vulnerabilities and decision making about health vary over time and space and is concerned with inequities between different social actors, groups, communities, and regions.

Human health is not a fixed concept but a fluid domain produced through the dynamic interactions between social and ecological systems. Health is not a state of being but an ideal to be achieved. Its inherent fluidity negates the possibility of a fixed state of health. Rather, health unfolds over time and space in highly varied ways that are simultaneously embodied by individuals who are located within broader structures of power that enable and constrain the possibilities for health and well-being. Given this understanding of health, what does it mean to identify a person as healthy? To consider someone "healthy" because she is free of a particular disease does not necessarily incorporate concerns about the quality of her environment and the amount of daily exposure to toxins and pollutants that affect her health over time. The physical body may be strong, but what must be said of an individual who suffers from a mental illness that diminishes his functioning and consumes the energies needed for a high quality of life? What of the decreased ability to pursue healthy activities with limited opportunities or resources? We are never fully healthy, in the same way that we are never solely diseased, and these fluid states of disease shape the possibilities for human health. My intention is therefore not to present a single definition of health, but rather to engage with the ways in which health is produced and the often highly unequal effects of that production. In centering the analysis on the generation of disease and health, I highlight variations in vulnerability and the divergent lifeways through which people are able to pursue their well-being.

This expansive understanding challenges approaches that tend to narrowly define health as the absence of disease. Thus, it requires a framework that does not reduce its spatial and temporal analysis to focus strictly on the individual body, but rather situates the individual in his or her broader context. Human health exists at the interface of environment and society; therefore, the linkages across spatial and temporal scales, the networks that produce vulnerability and healthy decision making, and the stability of social and ecological systems are all critical elements that factor into its production. As outlined in chapter 1, several academic disciplines and policy realms—including geography, social epidemiology, sociology, and public health—have made related contributions. But this book is novel in its demonstration of this perspective through the inclusion of several different health challenges in varied settings. This approach is needed because there remains a tendency to focus on disease in isolation from other health threats or the larger context. As I will argue throughout the book, part of the legacy of epidemiology—the field of biology that studies the factors causing the spread of particular diseases—has been the isolation of illness and the removal of the diseased subject from their political environmental context.

Chapter 1 outlines the social ecology of health framework in greater detail to demonstrate how human health is a fluid domain that is shaped by social systems, ecological systems, and their reciprocal interactions. I also present what I call "biomedical development," which is the coupling of biomedicine with socioeconomic development. In essence, biomedical development asserts that human health is largely a natural process that is depoliticized. This means that disease is spread through infectious organisms rather than the broader political and economic context. When the role of the state is taken into consideration for human health, it is generally unquestioned that economic development produces better health. This assumption has been challenged by studies in the social sciences, and I draw upon them in order to detail the ways in which spatial processes and the political environmental context produce inequitable health outcomes.

Chapter 2 analyzes the HIV/AIDS epidemic in South Africa. As the opening vignette demonstrated, the spread of HIV and the resulting national response were bound up with discursive battles. HIV/AIDS also created new social conflicts that were tied to deeply rooted patterns of racial segregation. More than twenty years after the democratic transition, the epidemic has shifted course, largely as a result of emerging discursive strategies, social-movement activism, and changed governmental responses. This encouraging development means that HIV is now a managed condition for many in the country, as long as they are tested and can access antiretroviral therapy. But as I argue in the chapter, managed HIV involves more than drugs. It is a lifeway that includes nutrition and caloric needs, access to ARVs, food security, and governmental documentation. Perceptions of health and well-being are interlinked with natural resource consumption and the use of traditional medicine. As such, HIV/AIDS also has ecological features.

Chapters 3 and 4 extend the analysis of South Africa's HIV/AIDS epidemic by connecting some of its contemporary patterns with historical systems of social and spatial regulation. I provide a history of colonial rule and the apartheid Bantustan system, focusing in particular on the creation of KaNgwane, to show how these spatial systems remain meaningful in shaping HIV landscapes for residents in the contemporary period. For example, racial segregation under British colonialism was facilitated through anxieties (many of them manufactured) over infectious disease. This has prompted some to assert that hostility to state-directed health interventions in South Africa should be understood from this history,[34] underscoring the sociopolitical dimensions of human health. The implications of the spread of disease and the potential for healthy lives are mediated by the governance regimes, institutional frameworks, gendered practices, cultural mores, and livelihood systems that exist within these territories. I argue that these historical spaces contribute to producing the landscapes of HIV, in ways that make certain social actors more vulnerable to infection and less able to access care than others. These historical landscapes are integral to the political environmental context of managed HIV.[35]

Chapter 5 details the roles of biophysical processes and socio-ecological change in contributing to human health. The IPCC Fifth Assessment Report (2014) indicates a number of pathways for climate change and health, emphasizing how the natural environment plays a formative role in shaping the spread of infectious disease and the potential for health management.[36] Increasing ecological variability, whether from precipitation or from flooding, is anticipated to affect human populations in multiple ways. Disease vectors are mobile and responsive to ecological change; therefore, even slight variations in temperature or other biophysical conditions can alter vulnerabilities. Changes in flooding dynamics can also reshape the possibilities for human health and well-being. Drawing upon research in the Boteti River region in northern Botswana, I argue that changing flood patterns contribute directly to how human health is understood. Additionally, opportunities for well-being are generated by the ways in which increased water affects individuals and families living next to the riverbed. Some residents are better positioned, whether financially or through social networks, to take advantage of the changes by increasing their income and food supplies, thereby improving the quality of life for their families. Others are less able to adjust to the changes, and their livelihood possibilities will be disrupted, to the detriment of their health decision making. Simply put, the return of the water in northern Botswana shows how the human health domain varies between different social actors and that these variations are produced by the political environmental context of the Boteti region.

Chapter 6 concludes *States of Disease* with a discussion of what these cases mean for understanding how the production of space shapes disease exposure and health–environment interactions, with the intention of supporting more holistic and interdisciplinary approaches that minimize the spread of disease and improve human health and well-being.

1 Social Ecology of Health

The announcement by the World Health Organization (WHO) in November 2008 that an estimated 6,072 people in Zimbabwe had been infected with cholera, and that 294 people had already died, received international attention and nearly unanimous condemnation.[1] What was particularly notable was the insistence by members of the global health community that the Zimbabwean government was responsible for the outbreak. Specifically, health experts attributed the epidemic to the poorly maintained sanitation services and lack of clean water in urban centers that allowed the disease to spread. Efforts to lay blame on the government dovetailed with long-standing pressure to remove President Robert Mugabe from power. As the number of infected people continued to increase in December, the government cut water supplies to Harare because it had run out of the chemicals necessary for treatment.[2] A public protest by doctors and nurses in Harare a day later was disrupted by Zimbabwean riot police, and an emergency response to the outbreak was announced by the United Nations Children's Fund (UNICEF), the European Commission, and the International Red Cross.

With an estimated 12,000 infected and more than 560 dead in December, David Parirenyatwa, the Zimbabwean health minister, declared

the cholera outbreak a national emergency and requested outside assistance. British Prime Minister Gordon Brown responded by calling the
cholera crisis an "international emergency" and urged the global community to take action. The responses by other international leaders were not
restricted to the cholera outbreak. One day later, Kenyan Prime Minister
Raila Odinga called for troops to "dislodge" President Mugabe. Other
world leaders, including U.S. President George W. Bush and French
President Nicolas Sarkozy, used the crisis to signal that Mugabe needed to
be removed from power. While the WHO announced that up to 60,000
people could be infected if the epidemic worsened, Zimbabwe's information minister, Sikhanyiso Ndlovu, called the situation "under control" and
declared that the West caused the crisis in order to facilitate a military
intervention.[3] Ndlovu specifically described the outbreak as a "genocidal
onslaught on the people of Zimbabwe by the British" and "a calculated,
racist, terrorist attack on Zimbabwe" intended to overthrow the Mugabe
regime.[4] The outbreak of cholera continued to intensify, and by July 2009
at least 4,288 people had died, out of an estimated 98,592 cases of infection. When the government of Zimbabwe then declared the epidemic to
be over, Peter Salama of UNICEF responded by stating that another outbreak was "almost inevitable."[5]

    This incident helps demonstrate that human health is generated by
social systems that contribute to shaping transmission patterns and the
ability of health-care agencies to respond effectively. A striking feature of
the cholera outbreak was its labeling by health-care practitioners as being
socially produced. In fact, the spread of this preventable disease led Dr.
Douglas Gwatidzo, head of the Zimbabwean Association of Doctors for
Human Rights, to state that the epidemic was "man-made."[6] The framing
of the outbreak as a social construction was supported by the infrastructure conditions, such as sanitation and provision of drinking water, that
had deteriorated during the economic recession of the Mugabe regime.
These factors contributed to an institutional and technical landscape in
which vulnerabilities to disease were generated. Additionally, the cholera
outbreak demonstrates that human health is understood and contested by
various social actors, with decided implications for the resulting political
responses. In the Zimbabwe case, this involved the positioning of cholera
within broader geopolitical considerations, including the effort to remove

Mugabe from office. As this situation reveals, disease discourses unfold upon a contested terrain because these representations are simultaneously political. If a disease is seen as being produced by poor sanitation, then economic development is understood as a logical response. It is widely understood that cholera results from exposure to the bacterium *Vibrio cholerae,* which occurs in certain settings through deficits in infrastructure and sanitation systems. Yet, at a different scale of analysis, if that same disease is presented as the product of direct exposure to an infectious agent or microorganism, then removal of contact is the recommended course of action. Both of these perspectives might be presented simultaneously to benefit certain agendas. And while these responses involve human populations and the modification of social landscapes, their trajectories are distinct and can serve competing interests.

It is important to emphasize that these representations are powerful because they generate responses from institutions that make targeted decisions about how scarce resources are to be allocated for disease testing, surveillance, monitoring, treatment, and future planning. Like other infectious or communicable diseases such as West Nile virus or HIV, cholera is framed in distinct ways that can underemphasize the underlying structural conditions that make people differentially vulnerable. Disease discourses are not monolithic or uniform; rather, they can vary between social actors, depending on individual understandings, perceptions, biases, educational training, and socioeconomic class. The ways in which a particular epidemic is perceived and negotiated between social groups results in a contested terrain that reveals underlying structural processes at work in a given time and place.[7] These disease landscapes are embedded in social systems that produce differential vulnerabilities and trigger distinct state responses and management regimes.[8] The social context, therefore, creates the conditions through which these events unfold and contribute to shaping disease discourses. The resulting responses can therefore be technical, and depoliticized, in design. As a result, they can be ineffective in ameliorating the underlying conditions or in preventing future outbreaks from occurring. This was reflected in Peter Salama's assertion that another outbreak was forthcoming, which proved correct when new cases of cholera appeared in May 2012 in the rural village of Chiredzi.

There remains a tendency in some of the scholarship on human health to downplay the role of structural sociopolitical factors. Instead, infectious disease tends to be treated as a natural event that occurs because of ecological conditions or behavioral decision making that produces vulnerability.[9] This is despite the fact that social dynamics can be determinative in shaping how disease spreads and which actors are able to access health care. These interactions take a variety of forms, depending on the site of analysis, but can include the built environment, spatial histories, state practices, understandings of disease and well-being, and ability to access and utilize health-care information and centers. All of these must be considered if we are to understand the possibilities for human health. Conversely, research on the social determinants of health can overlook ecological conditions by emphasizing social factors in the spread of an infectious disease or in producing vulnerabilities in human populations. While social conditions were blamed for the cholera outbreak in Zimbabwe, *Vibrio cholerae* survives in specific environments. Ecological factors contribute to the underlying spatial conditions through which the domain of human health either expands or contracts over time, thereby revealing differential disease vulnerabilities and unequal opportunities for health and well-being. It is the *social ecology of cholera* that produced Zimbabwe's particular outbreak and that contributes to the possibility of future occurrences.

BODIES OF DISEASE AND ILLNESS IN ISOLATION

To begin, it is necessary to engage with some of the dominant features of how human health is understood. I should highlight that these features are not monolithic within disciplines, as there are subfields that present competing views; however, for the purpose of clarifying the social ecology of health framework, I will outline three themes. First, it is necessary to examine how disease is traditionally understood within Western biomedicine. The biomedical model forms the centerpiece of the clinical medical tradition that has been in practice since the nineteenth century. It is derived from a transition in modern medical history whereby germ theory supplanted miasmatic theory, which posited that a disease like cholera or malaria was caused by pollution or a harmful form of air. Proponents of miasmatic theory asserted that diseases were produced by specific land-scape features, such as marshes and swamps that coincided with their incidence.[10] An outbreak of cholera in the Soho area of London in 1854 proved consequential in generating new understandings of the incidence and underlying causes of disease. Nearly five hundred cases of cholera occurred over a ten-day period in a concentrated area between Cambridge Street and Broad Street.[11] British physician John Snow, who was an early proponent of germ theory, conducted interviews of families that had been affected by the outbreak and created a map showing the spatial patterns of disease incidence. Also on his map were the locations of hand water pumps that, when taken in combination with the spatial distribution of cholera, allowed him to identify a particular pump on Broad Street as the source of the outbreak. In his detailed notes about the investigation, Snow wrote:

> On proceeding to the spot, I found that nearly all the deaths had taken place within a short distance of the [Broad Street] pump. There were only ten deaths in houses situated decidedly nearer to another street pump. In five of these cases the families of the deceased persons informed me that they always sent to the pump in Broad Street, as they preferred the water to that of the pumps which were nearer. In three other cases, the deceased were children who went to school near the pump in Broad Street.[12]

The removal of the Broad Street pump coincided with the recession of cholera cases, which made Snow a national celebrity and an early leader in the emerging field of epidemiology.[13] Snow also demonstrated the power of coupling disease-incidence mapping with qualitative interview-ing to uncover seemingly invisible layers of human decision making and social interactions. This showed that cholera vulnerability was not simply the product of geographic location, but was shaped by how social actors engaged with their environments in dynamic ways. By locating individual cases within a broader spatial context, Snow revealed the hidden land-scape of cholera that supported an effective public health response.

Snow's interventions were thus noteworthy and foreshadowed shifts in conventional wisdom about the causes of disease incidence. As noted by Kenneth Mayer and H. F. Pizer, "Humankind has been perpetually aware of the presence of invisible agents that could devastate communities with great rapidity. . . . Advances in microbiology enabled nineteenth-century investigators to identify etiological agents of communicable disease outbreaks and to develop more rational approaches to disease prevention and treatment."[14] This approach guides the field of epidemiology in studying the causes, patterns, and determinants of disease in human populations. The focus on distinct disease patterns is well established; it is based on standard approaches in epidemiology and germ theory that help outline the features of the biomedical model. The biomedical model focuses on the pathology, biochemistry, and physiology of a particular disease. Disease pathology studies the physical locations where disease exists, centering on the organs, bodily tissues, fluids, and entire bodily structure. Biochemistry examines the human body as an organism, focusing on the structure and function of cells and biomolecules. Physiology studies the mechanical, physical, and biochemical functions of living organisms. Taken together, these three elements serve as the foundation for the Western biomedical tradition that currently guides interventions into disease management and public health.

The biomedical model tends to center on the human body as the site of disease.[15] The consequence of this perspective is that acquired immunodeficiency syndrome (AIDS) is understood as developing from infection by HIV, which is transmitted through sexual activity, intravenous drug use, blood transfusions, or other activities that facilitate the transfer of bodily fluids. The cholera outbreak in Zimbabwe resulted from exposure to *Vibrio cholerae* through infected water supplies. Malaria, which causes morbidity or mortality in millions of people each year, is understood as being transmitted from *Anopheles* mosquitoes that carry the protozoan parasites in blood. These diseases are thus seen as the result of exposure to the virus, bacterium, or parasite, thereby centering attention and intervention on the human body.

A central issue with these particular disease framings is that they can narrow the scope, putting broader determinants of health outside the frame. For example, one of the lessons of Snow's study on the cholera outbreak was that where an individual is located matters, *but location is not spatially fixed.* Residents in the neighborhood were active agents and came into contact with the Broad Street pump through their daily movements and individual preferences. The distillation of a disease pattern to a static spatial dynamic fails to attend to the underlying social and eco-logical factors that contribute to both disease vulnerability and the possi-bilities for healthy decision making. Because of this, the biomedical model has been challenged for downplaying the role of social factors in disease transmission.[16] Wade and Halligan have argued that the biomedical model fails to explain many forms of illness because it assumes that "ill-ness has a single underlying cause, disease (pathology) is always the single cause, and removal or attenuation of the disease will result in a return to health."[17] This means that the multiple, interrelated, and holistic dimen-sions of human health are underemphasized. This is not meant to suggest that epidemiology and other fields are unaware of these dynamics, but simply to assert that the conceptual framework of biomedicine can restrict the application of such a perspective.

Second, a direct consequence of the biomedical model is that an indi-vidual's vulnerability to disease is not fully understood. As subsequent chapters will demonstrate, social and spatial processes intersect in pro-ducing the conditions that make certain populations vulnerable to disease while also enabling and constraining healthy decision making. These vul-nerabilities vary over time and have been historically created through the spatial production of health landscapes. In her work on infectious disease in Tanzania, Meredeth Turshen directly challenged the biomedical model for failing to provide a holistic analysis of the interaction of people with their economic, political, and social settings.[18] Turshen suggested that this stemmed from clinical medicine's focus on the individual rather than the collective, which failed to attend to the position of an individual in rela-tion to their larger environment. In a later study of disease in Tanzania under colonial and postcolonial states, she asserted that understanding health in Tanzania necessitated attention to colonial relationships and spatial patterns that were linked to political economic arrangements that advanced the power of particular social actors. She thereby provided a strong challenge to the biomedical model, explaining that the "Cartesian paradigm takes individual physiology (as contrasted with broader social conditions) as the norm for pathology and locates sickness in the indi-vidual's body (as opposed to the body politic)."[19]

One consequence of the biomedical framing is that diseases are seen in particular ways and at specific sites: the infectious agent and the individual body. Emphasizing the advancements of modern medicine derived from the biomedical tradition, Paul Farmer noted that "the narrow or uncritical use of these tools is one reason for physicians' blindness to the large-scale forces that generate sickness."[20] The narrowing of focus to particular diseases underemphasizes not only the underlying social and structural conditions that produce vulnerabilities, but also the opportunities and decision-making power of social actors to live healthy lives. While we certainly know something about the health of an individual diagnosed with malaria, we do not know the full gamut of lived experiences that make that individual differentially vulnerable. This is a dangerous absence, because it can assign much of the responsibility to that person while suggesting that behavioral change is the most effective way to prevent the occurrence of disease.[21]

Focusing on a particular disease agent or individual body tends to isolate one disease rather than address the multitude of factors and their relationships in producing the domain of human health. Syndemics is the study of the compound effects of more than one disease on the health of an individual or group, recognizing that diseases interact synergistically in producing vulnerabilities. These patterns tend to emerge within socially marginalized groups, with the implication that effective public health responses require attention to how their underlying causes reflect structures of inequality.[22] Research on syndemics has shown how diseases interact in intensifying health vulnerabilities for populations, while also assessing disease within its biocultural environment.[23] Syndemic interactions can range from the coincidence of HIV and tuberculosis, or of kidney disease and heart disease, and demonstrate how health vulnerabilities are constituted by interacting threats. For example, Eileen Stillwaggon detailed how HIV/AIDS is intensified by the exposure to parasitic and infectious diseases that weaken the immune system. Additionally, malnutrition undermines the immune system and makes individuals more susceptible to infectious diseases, including HIV.[24] Analyzing more than one disease or health threat can help identify linkages between them that could otherwise go unnoticed. While malaria and HIV are distinct viruses with specific etiologies, there are similarities in the social and ecological conditions that make certain individuals and populations more vulnerable to exposure.

Third, the particular disease framings resulting from the biomedical tradition generate specific health interventions. While Western biomedicine is generally effective at isolating diseases to the level of the germ and the site of the human body, this translates into particular forms of intervention for treatment and management. In other words, this focus is not without consequences. The resulting pattern has been demonstrated for decades within the fields of governance and international development, often drawing from Michel Foucault's theorization of "governmentality." In a richly detailed account, James Ferguson engaged with the ways in which development ideas gain traction through ideology and practice.[25] Central to his analysis was that the machinery of global development benefits from making poverty a technical challenge that can be simply addressed through the correct mix of knowledge and policy. Ferguson argued, however, that these interventions often fail because they are depoliticized, thereby missing the sociopolitical dynamics that create underdevelopment. Also notable is Timothy Mitchell's historical account of how the modern Egyptian state managed populations and landscapes through technocentric strategies that marginalized competing perspectives.[26] Similarly, Michael Goldman has chronicled the role of the World Bank in championing certain development projects, particularly large hydroelectric dams, through a discursive rationality that privileges technoscientific and economic knowledges.[27]

The narrow framing of human health becomes evident in the form of technical and apolitical interventions for its improvement. The identification of a problem is a starting point for contestations between those with competing worldviews, disciplinary trainings, and agendas. The identification of the problem produces particular framings for how it is understood. Human and nonhuman landscapes are reimagined by external entities through depoliticized narratives that justify the transformation of those landscapes, even asserting that it is in their best interest. These moments of encounter between development agent and development subject need not be seen as sinister. Tania Murray Li has convincingly shown that many agencies operate with the "will to improve" the lives of those touched by the project of international development.[28] But improving and being improved upon are two distinct positions of power. In the field of global health, one of the dominant strategies is the invocation of economic development for the improvement of human health.

EPIDEMIOLOGICAL TRANSITIONS
AND BIOMEDICAL DEVELOPMENT

One way in which human health can be narrowly construed is through its unproblematic and apolitical linking with economic development. The global development project following World War II was marked by a philosophy called "modernization theory," which categorized countries into a structured and linear development pathway. An architect of this strategy was President Kennedy's advisor Walt Rostow, who wrote in *The Stages of Economic Growth* that with the proper combination of political, economic, and cultural reforms, supposed "traditional societies" would adopt the characteristics of wealthier countries to reach a stage of high mass consumption.[29] These development stages were an idealized path, involving a linear trajectory that would be realized through increased manufacturing and industrial organization, investments in infrastructure, and the emergence of social elites. Nation-states were presented organically, as biological entities that needed to mature from an undeveloped to a developed stage through a process of evolution. Modernization theory asserted that with the proper guidance, these traditional societies would mature into fully developed countries. The concept of development, therefore, adopted a connotation of growth, change, or maturation that was facilitated by the labeling of certain countries as "developed" and others as "developing."

Joel Wainwright asserted that the concepts of development and nature are linked by their tendency to indicate a process whereby an entity advances from one stage to another.[30] In this sense, development is seen as the unfolding of something essential, as in "plant development" or "child development." Wainwright's focus is on disentangling capitalism from development, since their conjunction implies that capital is natural. His argument can be extended further to consider the deeply ingrained understandings of development itself. As with modernization theory in the 1950s and '60s, the term *development* became associated with growth, advancement, and maturation. Much like ideas of early childhood development, nation-states were placed on a spectrum from less to more advanced. With the proper education and management, the development process would enable these societies to develop into mature entities, ideally equipped to successfully compete in the global economy. In likening this process to adulthood, development is then inscribed as an organic and natural process that cannot be challenged. To question development would be akin to questioning natural law.

I emphasize the organic and naturalistic elements of development because global health and development institutions have increasingly identified human disease as a disruption to social and economic development. HIV is a virus not just to the individual body but also to the body politic. Disease destabilizes demographic systems, social networks, agricultural production, and resource-extraction industries that enable the machinery of global development. A feature of much of the scholarly writing on HIV/AIDS is the representation of the disease as a shock to households or livelihood systems, based on the tradition of identifying shocks to economic systems.[31] Drawing upon his rural livelihoods framework, Frank Ellis suggested that diseases, along with drought, floods, pests, and civil war, are shocks that can modify access to capital assets through social relationships, institutions, and organizations.[32] Unlike stresses, which are smaller, predictable, and more regular, shocks are understood to be large, unpredictable, and irregular disturbances that can destroy assets. As Ellis explained, "Loss of access rights to land, accident, sudden illness, death, and abandonment are all shocks with immediate effects on the livelihood viability of the individuals and households to whom they occur."[33] Vulnerability frameworks have emerged in recent years to consider the ways in which individuals and larger populations are exposed to hazards and the degree of resiliency within the system to withstand these shocks.

Research is also demonstrating the poverty–epidemic cycle in which poverty increases the spread of infectious disease that, in turn, contributes to producing socioeconomic poverty. The global development community has recognized the coupled relationships between economic development and environmental health. As evidence of this, the United Nations Millennium Development Goals emphasize health and environment challenges for the global population. Target 6 is designed to combat HIV/AIDS, malaria, and other diseases, identifying the need to expand access to treatment for HIV/AIDS while also altering the conditions that facilitate the spread of malaria. Target 7 indicates the importance of environmental sustainability for human health. In addition to addressing biodiversity loss and sustainable policy making, the target emphasizes reducing the proportion of the population living without sustainable access to safe drinking water and sanitation. The urban population residing in informal and slum conditions was also identified as a key environmental health factor.[34] According to the United Nations, 863 million people were estimated to be living in slums in 2012, compared to 650 million in 1990 and 760 million in 2000. The share of urban slum residents in the developing world declined from 39 percent in 2000 to 33 percent in 2012. More than 200 million of these people were able to gain access to improved water sources, sanitation facilities, or durable or less crowded housing, thereby exceeding the Millennium Development target.[35]

The tendency is to equate economic development with health, or in other words *prescribe development to the patient*. Much like poverty, economic development is presented as the solution to human disease by providing income, access to social services, education, medical care, and improved sanitation that seemingly reduce the spread of disease. As Meredeth Turshen stated, "More fundamental still than issues of management, increased resources, and the transfer of medical science and technology is the question of improved health as a natural result of economic development. Classical liberal economics places hope for better health in economic growth."[36] In some cases the potential for economic investments in the Global South to destabilize existing social systems (e.g., agrarian production practices), and to increase dependencies on imported products, is noted in the field of global health.[37] What goes unaddressed by the health-as-development narrative are the ways that economic modernization increases vulnerability to infectious disease and exposure to noninfectious disease.

The confidence in the inherent benefits of global development for human well-being is shared by the epidemiological transition model. Popularized in the 1970s, this model asserted that the primary health threats for poor countries were caused by infectious and communicable diseases. By comparison, wealthier countries were more likely to contend with noncommunicable diseases (NCDs), such as heart disease and cancer. The epidemiological transition model posited that as poor countries experienced economic development, they would generate resources to invest in health-care facilities and infrastructure, thereby reducing their burden from infectious disease. Paul Farmer and colleagues emphasize that the epidemiological transition model has several problems, including the fact that many NCDs have infectious etiologies. Additionally, while it continues to circulate widely, they suggest that the model needs to reevaluate the double burden of disease in low- and middle-income countries where populations are contending with both noncommunicable and communicable diseases. They conclude that NCDs cause more deaths globally than infectious diseases and are responsible for 60 percent of global mortality, 80 percent of which occurs in developing countries.[38] Lastly, they note that dominant framings of NCDs focus on particular lifestyle risk factors, such as excessive alcohol consumption and diet, and particular diseases, such as health diseases, lung diseases, cancer, and diabetes. Importantly, this obscures the burden of other noncommunicable diseases among the global poor, including the impacts from mental health disorders.[39]

There are parallels between modernization theory and the epidemiological model, particularly in the minimization of societal forces, universality, and embrace of a simple cause-and-effect explanation. In *The Stages of Economic Growth*, Rostow asserted that the proper combination of systemic reforms would enable traditional societies to adopt the characteristics of wealthier countries to reach a stage of high mass consumption. Yet for all its grand ideology, modernization theory was distressingly vague about its sociocultural prescriptions. In linking "development" to Western cultural conceptions, it minimized local context and avoided the messy politics that intrude upon socioeconomic decision making. Both modernization theory and the epidemiological model are universal blueprints that can be applied to countries, irrespective of their unique dynamics. In attending to a disease like HIV/AIDS, the epidemiological model concentrates on reducing HIV transmission patterns through educational programs and access to contraception, which are behavioral public-health campaigns supported through a well-funded health infrastructure. Similarly, health-as-development justifies economic modernization as a way of transforming cultural practices and increasing access to infrastructure and social services.

Or consider the demographic transition model, which is typically presented as an alternative to neo-Malthusian depictions of human population growth that circulated in the 1960s and '70s. The demographic transition model identified four stages: preindustrial, urbanizing/industrializing, mature industrial, and postindustrial. In the preindustrial stage, crude birth rates and crude death rates remain close to each other, keeping the population relatively level. During the urbanizing/industrializing stage, however, improvements in health-care delivery and medicines, coupled with investments in sanitation and infrastructure, bring a sharp drop in crude death rates. Crude birth rates remain roughly the same during this stage, prompting an increase in the population rate. During the mature industrial stage, crude death rates continue to decline, and it is theorized that socioeconomic development in the society brings the crude birth rates down slightly; however, the overall population continues to climb in an exponential J-curve. The contribution of the demographic transition model is its response to the alarms raised by neo-Malthusians such as Paul Ehrlich; specifically, the model suggests that the postindustrial stage results in the crude birth rates and crude death rates closely narrowing, thus bringing the population increase to a plateau. The fact that population growth rates in certain, although not all, industrialized capitalist economies have tended to follow this pattern has given the model greater validity. The central point is that the demographic transition model reifies the health-as-development narrative in uncritically accepting the benefits of economic modernization for human health. While it is assumed that access to improved drinking water and social services will reduce societal death rates, there is no mention of the negative consequences of industrialization for social and environmental systems.

A paradox of health-as-development is that economic modernization disrupts societies and ecosystems in ways that increase vulnerability to disease. The construction of hydroelectric dams in parts of Africa disrupts biophysical conditions and has made individuals and communities vulnerable to infectious disease, including schistosomiasis.[40] The expansion of chemical fertilizer and pesticide use around the world has increased the vulnerability of agrarian populations and ecosystems.[41] Clearly, improvements in medical technology and investments in safe drinking water reduce death rates, yet what should we make of long-term diseases produced by economic modernization, such as type 2 diabetes? Diabetes has been on the rise in the United States and other wealthier countries and is associated with particular diets and a sedentary lifestyle that are seen as the accouterments of "development." As the workforce moves from primarily agrarian employment activities into more fixed work such as information technologies or service industries, there are reductions in physical labor and caloric expenditures that contribute to obesity. Diets heavy on meats, processed foods, and high-fructose corn syrup increase the likelihood of long-term health complications. And what should be said about how economic modernization increases exposure to air and water pollutants and carcinogens that are seen as the unfortunate byproducts of economic development? The chemical factories, landfills, sanitation facilities, and other hidden links in the commodity chain of late-stage capitalism must exist somewhere—and, as the U.S. environmental justice movement has demonstrated, these are regularly sited next to poor and minority communities. These "sacrifice zones" exist because "low-income and minority populations, living adjacent to heavy industry and military bases, are required to make disproportionate health and economic sacrifices that more affluent people can avoid."[42] The simplistic linking of economic development with progress, and the assumed benefits for human health, need to be reconsidered in evaluating the possibilities for human well-being in the twenty-first century.

Development as the prescription for health faces inherent and unstable contradictions because of the increased vulnerability that comes from economic modernization. As William Adams has explained, development is "a two-edged sword, promising to hack away at the choking creepers of poverty, but at the same time bringing with it unrecognized, unregulated and often deeply hazardous change. Furthermore, the risks development creates are not distributed uniformly, but are concentrated in space and time."[43] These inherent contradictions merit our attention, as does the tendency of the international community to focus on particular diseases at the expense of others. As evidence of this, Farmer and colleagues detailed the "neglected" tropical diseases that do not command the same attention as the Big Three: AIDS, tuberculosis, and malaria. While these diseases, such as schistosomiasis, hookworm infection, and ascariasis affect millions, funding for new treatments is limited because of market disincentives. They noted that relying on the marketplace to invest in new drug discoveries is dangerous because "drug development is demand-based, not need-based."[44] The consequence is that the biomedical model and the promotion of the health-as-development narrative limit how diseases are understood, experienced, and managed.

HEALTH JUSTICE

The health-as-development narrative is directly challenged when the out-
puts of economic modernization are shown to produce noninfectious disease in minority communities in the United States and other settings. In
essence, while economic development is identified as a cure for many illnesses, it can simultaneously increase vulnerabilities to new diseases and
exposure to toxins and carcinogens. The U.S. environmental justice movement generally credits its origins to a series of cases in the late 1970s and
early '80s that demonstrated how certain facilities, such as power plants,
landfills, incinerators, and chemical facilities, are often situated in specific
communities and regions. Environmental justice in the United States has
concentrated on spatial patterns of pollution and disproportionate health
effects to show that differences exist by race, ethnicity, socioeconomic
class, and location, whether in a rural or urban setting. Yet it was not until
a series of cases brought national attention to differential exposure to
environmental pollutants that a national movement emerged.

One such case that received widespread media attention was Love
Canal. Located in a middle-class, largely white community in upstate New
York, the leakage from an industrial landfill resulted in a nightmare for
residents. The Love Canal community was built upon 21,000 tons of toxic
waste that had been dumped by Occidental Petroleum Corporation, which
was previously Hooker Chemical Company. The company had managed
the landfill for three decades, which it then covered with soil and sold to
the city for one dollar.[45] The landfill became public knowledge when
record rainfall caused the improperly stored containers to leach "their
contents into the backyards and basements of 100 homes and a public
school built on the banks of the canal."[46] This was the cause of immediate
concern because the landfill was believed to contain eighty-two different
compounds, eleven of which were suspected carcinogens at the time,
including benzene.[47] Upon learning that her son's elementary school had
been built on top of the site, one resident, Lois Gibbs, became a community leader in pressuring the state and federal government for support.
This was made all the more necessary because the first response was to
relocate only those residents in immediate proximity to the landfill.
A 1980 U.S. Environmental Protection Agency (EPA) study, which was

leaked to the media, asserted that younger residents and schoolchildren evidenced an unusually high number of chromosomal breakages and higher risk of cancer and birth defects. This provided enough pressure for the national government to purchase the remaining properties and allow residents to relocate. The Love Canal Homeowners Association helped evacuate hundreds of families while the cleanup was undertaken. This case generated widespread attention to the horrors of carcinogenic exposure and led to the creation of the EPA's Superfund legislation, which prioritizes toxic waste sites for remediation.

The story of Love Canal continued in the years following the initial relocation of community residents. Thirty years after the discovery of the exposure, a follow-up study from the New York State Department of Health reported that residents, particularly children who had lived in the area, remained at higher risk for kidney, bladder, and lung cancers.[48] In 1988, the area was identified as restored, to the extent that new residents were able to buy homes below market price. Protests followed the announcement that the area was secure, and activists warned of a future health crisis. In 2011, a city crew repairing a sewer line found toxic waste less than half a mile from the remediated landfill site. State and local agencies asserted that this was residual material from the original cleanup; however, the find sparked widespread concern in the community, which resulted in more than a dozen lawsuits from families living near the site and also from previous residents seeking monetary damages for their exposure to these chemicals.[49]

While Love Canal remains a touchstone in the emerging awareness of the relationships between health, environment, and justice, a series of waste-facility sitings in minority communities provided the catalyst for the larger movement. In 1978, it was announced that a landfill to store polychlorinated biphenyls (PCBs) was to be placed in the Afton community in Warren County, North Carolina. The PCBs were the result of an illegal dumping incident between June and August 1978 by Robert J. Burns in collaboration with Robert Ward of the Ward PCB Transformer Company. Burns was later jailed for dumping the PCB-contaminated oil along the road in fourteen counties in North Carolina. The decision to place the landfill near the Afton community was controversial from the start because the site was in an area with high socioeconomic poverty and a predominantly African American population. Community members fought

Governor James B. Hunt's decision through litigation alleging that the site was not feasible because of soil permeability properties and groundwater levels. When the state began transporting more than six thousand truck-loads of the PCB-contaminated soil to the landfill in September 1982, it provoked widespread community resistance, including collective nonviolent direct action. During the six-week trucking opposition, Warren County citizens mounted a campaign of civil disobedience that received national attention and included more than 550 arrests. Ultimately, the organizers were unable to stop the opening of the landfill, though they continued to pressure Governor Hunt to honor his pledge that once technology became available, the state would detoxify the landfill. This was finally completed more than twenty years later. As a result of these efforts, the Warren County landfill is widely recognized as catalyzing the national environmental justice movement,[50] and as "the watershed event that led to the environmental equity movement of the 1980s."[51]

A similar decision to site a landfill in a predominantly African American neighborhood in Houston, Texas, expanded national attention to the dynamics between spatial segregation and political processes about where to locate landfills, incinerators, and industrial factories. The siting of Whispering Pines Landfill in a middle-class section of the city was particularly notable because of a lawsuit brought forward using civil rights legislation. *Bean v. Southwestern Management Corp.* was the first lawsuit in the United States that argued environmental discrimination in waste-facility siting under the Civil Rights Act.[52] The suit, which was initiated by Robert Bullard and his colleagues at the University of Houston, employed empirical data to demonstrate how different socioeconomic and racial groups faced disproportionate exposure to environmental pollution. As noted in a history of the lawsuit, in 1950 two-thirds of Houston's African American population was concentrated in three major, segregated neighborhoods.[53] By 1980, Houston's African American population had become further decentralized and occupied the northeast, northwest, southeast, and southwest corridors. While African Americans made up 25 percent of the city's population at that time, all five city-owned landfills and six of the eight city-owned incinerators were in African American neighborhoods. From 1920 to 1978, eleven of the thirteen city-owned landfills and incinerators were built in the same neighborhoods. After facing temporary injunctions, the legal case went to trial in 1984 and was ultimately unsuccessful. However, following the trial, the Texas Department of Health updated its permit requirements to include more detailed land-use information, including demographic detail, in hopes that this would make future siting decisions more equitable.[54]

The manipulation of space and political ordinances enabled the city to identify this neighborhood as the ideal location for the Whispering Pines landfill. Calling this a product not of classism but of "slam-dunk, in-your-face environmental racism," Bullard emphasized the absence of zoning as playing a central role in the placement of the facility.[55] Specifically, the city relied on deed restrictions that regulated lot sizes; square footage of structures; the distance that structures must be set back from property lines, street lines, or lot lines; the type and number of structures that may be built on a lot; and whether single-family or multifamily housing may be built there. The consequence of these restrictions is that the population composition of a neighborhood can be shaped in particular ways, which has consequences for the ability of people to mobilize to protest should a facility be sited in their neighborhood. Bullard has written about the ways in which racism and racial preferences are not necessarily explicit and intentional but can be structural and institutionalized. In subsequent chapters, I will detail the ways in which the South African state formalized social and spatial configurations in overt ways to benefit the white minority. These actions were accompanied by entitlement systems that became entrenched over time. Bullard emphasized that this has occurred in the United States in a similar fashion, noting that

> environmental racism buttressed the exploitation of land, people and the natural environment. It operates as an intra-nation power arrangement— especially where ethnic or racial groups form a political and/or numerical minority. For example, blacks in the United States form both a political and numerical racial minority. On the other hand, blacks in South Africa, under apartheid, constituted a political minority and numerical majority. American and South African apartheid had devastating environmental impacts on blacks.[56]

Justice movements have shown how human health is constructed out of spatial dynamics that unfold over time through places and landscapes.

Three examples of how spatial processes shape the political environmental context for urban and rural landscapes are redlining, food deserts, and urban green space. In 1935, the Federal Home Loan Bank Board tasked the Home Owners' Loan Corporation (HOLC) with evaluating 239 cities in an effort to generate "residential security maps" to prioritize areas for future real-estate investment. HOLC was established during the Great Depression and was originally designed to reduce residential foreclosures, and between 1933 and 1936 it provided more than a million loans to those at risk of default.[57] Figure 1 is a representation of HOLC's 1936 map of Philadelphia.

The areas designated as having either the "best" or "still desirable" housing stock (in green and blue, respectively, on the original map) were seen as warranting future development. "Hazardous" areas (red in the original) were targeted for reduced investment. These maps are credited with the origins of the term *redlining*, which is the steering of economic resources into certain regions and not others, primarily because of the racial and ethnic composition of the resident population. While there have been challenges to the assertion that these maps caused specific instances of redlining, Hillier reviewed them using various techniques and confirmed that the racial composition of these neighborhoods was a determining factor in the poor grades.[58] These maps, and the political and economic processes they generated, provided a spatial foundation for future investment in ways that contributed to political and economic inequities in urban landscapes in the United States.

The concept of a "food desert" is widely utilized in academic and public health communities to refer to geographic territories that lack access to high-quality and healthful foods. In certain urban neighborhoods, it is not possible to find a full-service grocery store that offers fresh fruits and vegetables for sale. Rather, convenience stores predominate. Similarly, in rural areas, there can be a shortage of grocery stores that offer specific types of commodities, including many of those recommended for good health. For populations that lack transportation, this can result in limited or no access to these types of items. In cases where transportation is available, it can mean that purchases that take minutes for some can take hours out of the week for others. In a review of some of the existing literature on food deserts in the United States, Renee Walker and colleagues

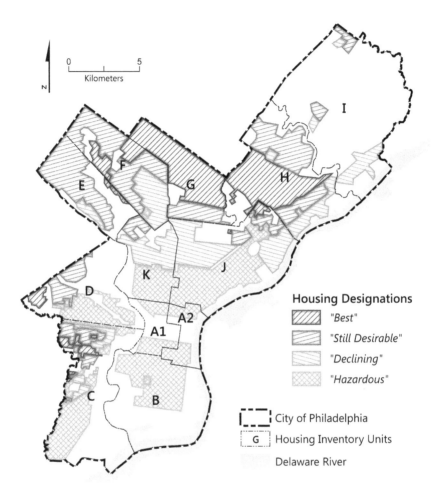

Figure 1. Representation of the Home Owners' Loan Corporation map of Philadelphia (1936).

found four dominant themes.[59] First, access to supermarkets varies widely. In Philadelphia, one study showed that the city had the second-lowest number of supermarkets per capita compared to other major U.S. cities in the 1990s.[60] Second, the studies showed that racial and ethnic factors play a role in the availability of grocery stores. Third, socioeconomic poverty influences the size of grocery stores and the services they offer. And fourth, there are differences in the prevalence of chain and non-chain stores. The predominance of smaller grocery stores and fewer supermarkets has the potential to drive up prices, making it more difficult for economically-insecure families to maintain food security. Larger supermarkets have the capacity to sell both brand-name and generic items in a variety of package sizes, which can offset higher prices and support the purchasing power of people living at or below the poverty line.[61]

The food desert concept has limitations. In many assessments of urban areas and food access, there is an overreliance on full-service grocery stores as the ideal entity for providing healthful food. Even in the absence of these stores, in some settings neighborhood residents have access to vegetables and fruits through small shops, food pantries, or community gardens that are overlooked by a narrow framing of what constitutes a food desert. Additionally, research has shown that locating larger stores in certain neighborhoods does not invariably change resident preferences.[62] Regardless of these challenges, the concept of a food desert has been effective in coupling differential access to certain types of services with the spatial dynamics that produce urban and rural landscapes. These landscapes involve gendered, generational, cultural, and political dynamics that challenge location and proximity as the sole metrics shaping community health and well-being. They also reveal that decision making around consumption is constrained by spatial patterns that shape the political environmental context for food access and security.

While at its origins the environmental justice movement in the United States tended to concentrate on pollution and proximity to facilities such as landfills and toxic factories, the justice concept has broadened to consider differential access and proximity to factors that contribute positively to human health and well-being. Environmental justice is the idea that all people should have equitable access to environmental benefits and protection from environmental harms, regardless of race, gender, income, or other characteristics. In urban areas, minorities and economically disadvantaged populations are often disproportionately exposed to environmental burdens such as pollution and are less likely to have access to environmental benefits, such as parks, trees, and green space.[63] Soils contaminated with lead present a high health risk to children, and African American children between the ages of one and five are more likely to experience lead poisoning than white children in the same age bracket.[64]

While proximity to environmental hazards is unequally experienced, so is access to green space that provides opportunities for recreation, exercise, and contemplation. These variations are similarly produced by spatial processes that generate disparities for populations, with direct consequences for the places and landscapes of health.

PLACES AND LANDSCAPES OF HEALTH

Geographic research on human disease has been concentrated within the subfield of medical geography, which has focused on the spatial and ecological dimensions of human disease and health-care delivery.[65] The field of disease ecology has been foundational in demonstrating the complex relationships between humans and the natural environment in producing disease.[66] As Jonathan Mayer has explained, disease ecology examines "how humanity, including culture, society and behavior; the physical world, including topography, vegetation and climate; and biology, including vector and pathogen ecology, interact together in an evolving and interactive system, to produce foci of disease."[67] In a review of the field, Wilbert Gesler suggested that medical geography has expanded into several new areas since the 1980s, including studies on the distribution of health services, health inequalities, and the relationships between gender and disease.[68] This assessment dovetailed with the emergence of health geography, which to some scholars represents a distinct field from medical geography "indicative of a distancing from concerns with disease and the interests of the medical world in favour of an increased interest in well-being and broader social models of health and health care."[69] This "decentering" of the medical draws upon insights from social and cultural geography to engage with new epistemologies of health while attending to overlooked subject matter such as disability or sexuality.[70] The result has been a broadening of the concept of health beyond the absence of disease to consider the relationship of people to their environments in addition to their physical and emotional well-being.

The expansion of health geography has resulted in exciting engagements with the places and landscapes of human health. Robin Kearns and Graham Moon explained that what defined this "new health geography" was the use of place for understanding health, the application of social theory, and critical perspectives on the geographies of health. Additionally, landscape served as "a metaphor for the complex layerings of history, social structure and built environment that converge in particular places."[71] The concept of a "therapeutic landscape" has also received interest within health geography, and Fiona Smyth has argued that therapeutic networks between social actors, many of which exist outside of the biomedical model, necessitate attention.[72] *Place* has been described as "one of the most multi-layered and multi-purpose words in our language."[73] It is understood not as a location or portion of geographical space, but as being constructed and reconstructed out of a particular set of social relations, experiences, and understandings.[74] In addition, place is generally understood as a dynamic process that is constantly shaping social realms, while also being reordered by the individuals in those spheres. For example, Donald Moore analyzed conflicts among a differentiated peasantry and state in Zimbabwe to show how the landscape was a materially and symbolically contested terrain.[75] Moore was particularly interested in place-based politics, explaining:

> While the meanings of place are politically contentious, so too are the *practices* that carve out those meanings, rework them, and enunciate their form through historical struggle. Where these practices unfold, how they are mapped in social memories, and their influence on popular understandings of the relationship between identity and locality, all constitute a politics of place.[76]

The political environmental context contributes to the lived experiences of individuals, families, social groups, and populations as they navigate through places and landscapes that have been constructed over time. As Antony Cheng and colleagues stated: "By taking a place perspective, one recognizes that human connections with natural resources and the landscapes in which they occur are multifaceted, complex, and saturated with meaning."[77] In South Africa, the landscapes of colonialism and apartheid were produced over centuries through the aggressive management of human populations and spatial economies. This history has contributed to the infrastructure and health-care facilities that are available for local populations to access testing and life-saving drugs. The possibilities for human health are shaped by the availability of these services at sites that operate as a central point of contact between the state and its citizens.

If human health is shaped by the availability of services provided in part by the state, then, even in a narrow sense, health is inherently political. The field of political ecology is helpful for understanding the politics of place, landscape, and human health. Political ecology generally concentrates on the interactions between political economy and environmental resource use through the application of a scalar approach that draws links between various actors to analyze the contextual realities of decision making. Paul Robbins has argued that political ecology is a "text" that is constituted around five key theses.[78] First, the thesis of *degradation and marginalization* examines the ways in which forms of economic production and natural resource extraction become concentrated and unsustainable with greater integration within regional and global markets. Traditional property arrangements are destabilized by state and market forces that can have repercussions for local livelihood systems in ways that make them less equitable. Second, the thesis of *conservation and control* has emphasized the ways in which colonial and postcolonial governments utilized discourses of biodiversity protection to reframe territories through enclosures and dispossession. Third, political ecology scholarship has attended to the *environmental conflicts* engendered between social groups by access to and control over natural resources. This resonates with the fourth thesis, in which competition and control over natural resources produce *environmental subjects and identities* that are constructed through relationships with state, development, and corporate entities. Robbins suggests that new forms of environmental actions and institutional systems produce new kinds of people that perform as environmental subjects. And fifth, political ecology considers nonhuman nature as *political objects and actors* that shape the world around them, including human societies. This provides an avenue for addressing ecological systems and entities as agents that influence social systems in meaningful ways.

Political ecology scholarship on human disease and health has unfolded in two fairly discrete waves.[79] The first phase was primarily based in medical anthropology and medical geography and tended to concentrate on disease ecologies and the places of health. Central to this first phase was the work of Jonathan Mayer, an epidemiologist situated in the medical geography and disease ecology traditions. In a widely cited paper, Mayer advocated for a "political ecology of disease approach" to demonstrate "how large-scale social, economic and political influences help to shape the structures and events of local areas."[80] This generated a wave of scholarship that concentrated on disease patterns and health vulnerabilities as opposed to more holistic perspectives on human health. The second phase of political ecology scholarship tends to approach health in broader terms, addressing it less as the absence of disease and more in terms of long-term well-being and individual agency. One defining feature of this work is the commitment to examining conceptions of health and environment, and how they are connected with socioeconomic class, race, gender, and ethnicity. This scholarship has also engaged with a diverse array of subjects, such as embodiment, epigenetics, and socio-ecological production to interrogate disease vulnerabilities and the lived experiences of those attending to disease and managing their health.[81]

SOCIAL ECOLOGY OF HEALTH

Human health exists at the nexus of social and ecological systems and is produced through spatial patterns that are multifaceted and dynamic.[82] A continuing challenge to effectively addressing human health, and the underlying conditions shaping it, is that these conditions can be vast and varied. Research on the social determinants of health draws upon a number of disciplines and demonstrates the role of culture, gender, political economy, land-use patterns, religion, and other social factors in making human populations vulnerable to disease and proximate to conditions that generate poor health. For example, the HIV/AIDS epidemic intersects with social systems and is produced by them, in numerous ways. Employment patterns in industrial sectors, including mining, have been documented to contribute to the transmission of HIV. Drawing upon work from a gold mine in South Africa, Catherine Campbell detailed how vulnerability to HIV transmission is interlinked with masculinity and with the daily threats to survival from their occupation.[83] Miners interviewed in the study attested to the difficult work conditions and few opportunities for intimacy as contributing to risky decision making about sexual activity and condom use. Additionally, social inequalities in income and employment status tend to be associated with greater exposure to sexual activity, diminished access to health information, and delayed diagnosis or treatment.[84] Emphasizing a "political economy of risk," Farmer argued that "structural violence means that some women are, from the outset, at high risk of HIV infection, while other women are shielded from risk." He also noted that "women have been rendered vulnerable to AIDS through *social* processes—that is, through the economic, political, and cultural forces that can be shown to shape the dynamics of HIV transmission."[85] Farmer is insistent that socioeconomic poverty, race, class, and gender intersect in making certain social actors vulnerable to disease while others are protected from exposure.

The field of ecology examines the relationships among biological organisms and with their physical environment.[86] Shifts in temperature or rainfall, for example, can change transmission patterns to make some populations more vulnerable to infectious disease. One such disease is West Nile virus, which has been on the increase in the southern United States as a result of changes in temperature gradients that support the presence of disease-carrying mosquitoes.[87] As I will detail in chapter 5, ecological processes disrupted by climate change are challenging human health through the spread of infectious diseases into new areas. That chapter also demonstrates how increasing water in the Okavango Delta of northern Botswana is triggering livelihood adjustments that create possibilities for improved human health. Social systems influence the health domain and can be quite dynamic. Much like temperature increases, or changing levels of flooding, the construction of infrastructure or new forms of agricultural production can rework disease vectors, in some cases reducing transmission patterns in populations, as seen in the eradication of malaria in certain regions of the world. Housing foreclosures that leave lots vacant increase the amount of standing water that provides breeding habitat for mosquitoes that may carry West Nile virus.[88]

The social ecology of health framework asserts that the domain of human health is the result of spatial processes that are shaped by the coupled and dynamic intersections between social and ecological systems. Some of these conditions affect human health in dramatic ways, while in other settings they can be less determinative. The impacts also vary by

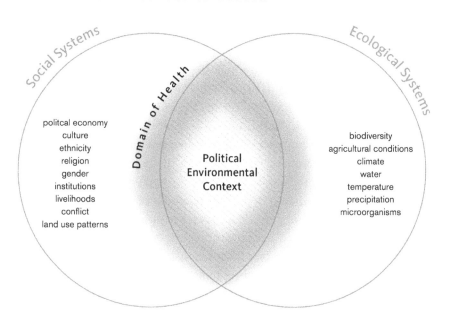

Figure 2. Social ecology of health.

particular diseases or the ways in which built environments contribute to making people vulnerable to poor health, such as the absence of green space for recreational purposes. Human health therefore exists as the domain, represented by fluid lines in the center of figure 2, that includes social and ecological systems and their interactions. The health domain is fluid because it varies depending on the particular frame of study—whether it is at the level of the individual or the population, in countries or across entire regions. Additionally, the domain of human health can expand or contract over time, depending on the current state of social and ecological systems and whether they are experiencing change. This approach to conceptualizing human health accepts that change is inevitable but that things can also improve. This can occur through state intervention or prevention programs and can be disrupted by development projects or ecological disturbances that modify vulnerabilities. While the states of disease are often presented as the product of a natural event that disrupts water and sanitation systems, the ways in which social systems are organized has proved critical in mitigating the severity of an outbreak. Human health also changes over time within social systems destabilized by climate change, including agricultural production and systems that provide water or other resources that contribute to increasing health vulnerabilities.

THE POLITICAL ENVIRONMENTAL CONTEXT

While I am concerned with human health in the fullest sense, my focus here is on the production of space and the particular elements that emerge at the interface between social and ecological systems.[89] I refer to this throughout the book as the "political environmental context," which is located in the center of figure 2. The political environmental context shapes human health by producing vulnerabilities to disease and the conditions that produce poor health. This context mediates the ways in which particular diseases are understood and managed while also enabling and constraining health decision making. Because the political environmental context changes over time, it can be unequal in terms of who is exposed to health threats and the conditions that undermine well-being. It produces inequities in determining which social actors are healthy or able to more effectively manage an existing illness through differential access to the best facilities, technologies, and medical practitioners. Simply put, the political environmental context creates the states of disease for individuals, families, and communities.

This focus on how the political environmental context produces human health makes three contributions. First, it attends to how historical spatial formations contribute to shaping contemporary vulnerabilities to disease or differential exposure to the factors that produce poor health. These patterns are rooted in political economic systems that unfold over time and space in often subtle ways. In some cases these benefits, whether the acquisition of material resources or the exercise of decision making and power, might be at the expense of other social actors. The political environmental context results in differential exposure to carcinogens and pollutants that generate chronic disease conditions for people who are often socially and economically disadvantaged or who are members of certain racial and ethnic populations. This has been an object of interest within some fields that attend to human health; however, the role of space in enabling and constraining these outcomes has been underemphasized. Space has been used as a mechanism for social control, whether in the apartheid policies of South Africa or the differential exposure to carcinogens in minority communities in the United States. While scholarship in the social sciences has been insistent on the structural conditions that produce poor health, it has not fully attended to how spatial processes produce structural constraints and inequitable opportunities for health and well-being.

Second, I consider health as a material but also a discursive formation. This means that while human health is tangible, it is also situational, relational, contingent, and dynamic. The consequence is that these factors result in multiple health outcomes within the same setting. As such, it is not just the places in which people are located that result in equal health outcomes; rather, the confluence of social and ecological circumstances produces the states of disease. Whether it is HIV/AIDS, malaria, or exposure to carcinogens, health challenges are distinct and particular to their context. However, they share a number of elements that require a comparative examination to properly understand human health. Addressing human health as a discursive formation means embracing often competing understandings of disease and well-being, because these varied perceptions are critical in exposing the factors that produce human health and inequities within societies. While a number of disciplinary domains recognize the role of these variables in human health, they are rarely integrated in an interdisciplinary and holistic manner.

Third, the political environmental context integrates ecological processes to consider how the natural environment is meaningful in the production of human health and well-being. Within the social sciences, it is more common to talk of "coproduction" between humans and nonhuman species, to give greater primacy to the mosquito or other vectors in spreading disease. Recent work has been intent on destabilizing the unidirectional axiom that nature is something transformed by humans, to consider the ways in which nature simultaneously transforms humans. The concept of "environmental health" refers to the surrounding environment that is produced by human beings but that should also be seen as something that produces humans. The WHO argues that environmental health addresses the

> physical, chemical, and biological factors external to a person, and all the related factors impacting behaviours. It encompasses the assessment and control of those environmental factors that can potentially affect health. It is targeted towards preventing disease and creating health-supportive environments. This definition excludes behaviour not related to environment, as well as behaviour related to the social and cultural environment, and genetics.[90]

Expanding upon this conceptualization of environmental health, the political environmental context gives agency to the natural environment so that it is understood as playing a direct role in shaping human health. Human health is therefore coproduced through the interactions between social and ecological systems.

I recognize that this theoretical engagement does not lend itself to a simple explanation for the presence and persistence of specific health challenges. Nor does it make for an easy fundraising campaign to eradicate a preventable disease. But that is not my intention. Rather, I seek to explain the underlying reasons that certain people are more vulnerable to disease while others are more likely to lead healthy lives. The social ecology of health framework obligates a more expansive and holistic perspective on the social and ecological determinants of human health and well-being. Additionally, it helps demonstrate how a particular disease, such as HIV, is differentially experienced. It helps explain why the same virus results in a distinct epidemic.

The points of connection between human populations and their surrounding environments in producing differential health outcomes are made clear when considering the HIV/AIDS epidemic in South Africa. The dramatic spread of HIV in the late 1990s and early 2000s exposed underlying structural conditions that made certain groups and regions more vulnerable to the disease and more challenged in managing it. The government's initial halting response, including resisting the rollout of ARVs, meant that vulnerabilities were socially produced in particular ways, as compared with other countries on the continent. The lived experiences of those infected and affected by HIV have been shaped by the political environmental context that involves not only social dynamics in the country, but also historical spatial patterns that have been produced to benefit segments of the population. While the epidemic has shifted course in recent years, largely as a result of more aggressive responses by the government and increased access to ARVs, the states of disease have significant implications for social and ecological systems and for the possibility of healthy lifeways in the era of managed HIV.

2 HIV Lifeways

I returned to Mpumalanga Province in November 2012 to continue research on livelihood systems, health–environment interactions, and conservation planning. Six years had passed since my last visit, and I was intent on assessing whether the landscapes of HIV had changed following President Mbeki's replacement by President Zuma and the departure of the health minister, Dr. Manto Tshabalala-Msimang. In fact, only weeks before my arrival, the recently appointed CEO of the South African National AIDS Council (SANAC), Fareed Abdullah, announced that the country had achieved universal access to HIV treatment.[1] SANAC presented new figures indicating that two million South Africans were currently on antiretroviral drugs (ARVs), which covered roughly 80 percent of those estimated to be in need of treatment. Abdullah stated that because universal access had been achieved, it was now important to increase prevention efforts. According to the report, the South African government funded about 80 percent of its HIV response domestically, with about 70 percent of this identified for treatment, leaving much less for prevention. In the beginning of 2012, the country met 30 percent of its condom distribution target by handing out 84 million condoms. The coordination of these activities was intended to occur across many levels, with national and provincial governments expected to cooperate in attending to the epidemic. South African Deputy President Kgalema Motlanthe, who chaired SANAC at the time, noted the importance of reinvesting in provincial AIDS councils, emphasizing that "each province has unique economic, social, infrastructural and cultural characteristics. There are population dynamics, health and community systems, and human resource issues that determine the impact of programmes, thus requiring context-specific approaches . . . for success."[2]

The announcement of universal access to ARVs was remarkable and circulated on IRIN PlusNews, which covers HIV/AIDS reporting from around the world. Universal access has been the goal of multiple health organizations, including the WHO, which published the report "Towards Universal Access" in 2010.[3] The report estimated that 34 million (range: 30.9–36.9 million) people at that time were living with HIV and that nearly 30 million (range: 25–33 million) people had died of AIDS-related causes since the first case of AIDS was reported on June 5, 1981. About 6.6 million people were receiving antiretroviral therapy (ART) in low- and middle-income countries at the end of 2010, representing a nearly twenty-two-fold increase since 2001. ART typically involves three drugs that suppress the replication of HIV and reduce the possibility of the virus developing resistance. For patients diagnosed with HIV, use of ARVs can mean that they remain at the clinical latency stage for decades, such that greater access to these drugs in the developing world ensures the possibility of long-term survival. Globally, a record 1.4 million people started life-saving treatment in 2010, which was more than in any previous year. According to the report, at least 420,000 children were receiving ART at the end of 2010, constituting more than a 50 percent increase since 2008.[4]

These achievements in increasing access to life-saving drugs are laudable. Yet the report also detailed the intercountry variations in their availability. According to the 2010 report, the proportion of people in South Africa that had access to ARVs in relation to the estimated number of people in need of treatment was 37 percent.[5] This was compared to estimates of 39 percent in Uganda, 48 percent in Kenya, 64 percent in Zambia, 76 percent in Namibia, and 83 percent in Botswana. It should be emphasized that these data were based on the 2010 WHO recommendations, which changed the advised protocol for administering ART from a CD4+ T cell count of below 200 cells/mm^3 to that of 350 cells/mm^3 or below (CD4+ T cells are specific blood cells that are necessary in helping the body fight disease). The impetus behind this recommendation was to get individuals on ARVs sooner, before the onset of severe symptoms, and to reduce the possibility of transmission to others. These new guidelines had the effect of increasing the number of people estimated to be in need of treatment in low- and middle-income countries by 45 percent. The WHO changed its recommendation in 2013 from a CD4+ T cell count of 350 cells/mm^3 to 500 cells/mm^3, meaning that an estimated one million additional South Africans could receive ART, thereby putting additional strain on the health-care sector.[6] In late 2015, the WHO announced that ART should begin immediately upon diagnosis of HIV, and anyone at risk of becoming infected should be given protective doses of similar drugs. Immediate treatment is the typical protocol in the United States and other developed countries, and extending this response to the Global South could mean that as many as nine million more people would require treatment.[7] It seems that while universalizing access is a laudable goal, it is also a moving target.

Given the challenges to universalizing access, it was surprising that members of the national government were announcing the achievement of such an ambitious goal less than three years after the WHO report was published. In order to investigate the assertion further, in November 2012, I spoke with a director of a local facility who had worked for years with home-based care workers. I asked if she had heard of this announcement, and she responded that she had not and appeared surprised by it. When questioned about what she thought of the government's assertion, she wryly remarked that it depended on your *definition* of universal access. During that conversation, she was insistent that the rates of HIV infection in the area were much higher than those being reported by the national government. Referencing a study of farm workers in Mpumalanga where HIV testing had been conducted, she suggested that there was an estimated infection rate in the range of 45–50 percent. This was consistent with what other NGO representatives were reporting to me in 2006, though I have never been able to find concrete verification of this number. I heard this estimate then and have continued to hear it since. It is elusive, yet powerful in its effect. The divides between national estimates and local understandings are stark and speak less to the "actual" numbers and more to conflicting perceptions of HIV's presence in the landscape. One way to interpret this would be to see the differences as simple inconsistencies between different social actors. Biomedicine and population-level statistics are open to interpretation and can be represented to serve particular agendas or molded to align with preexisting assumptions. Given the history of HIV/AIDS in this region, something much more complex, and political, is occurring. In employing these population statistics, local actors are speaking to larger institutions that frame and intervene in human lives. As such, they represent vocalized frustrations with the national government, in essence serving as a direct challenge to the discourse around HIV prevalence and declarations of its prevention achievements. These are statements of discontent with the messages conveyed about the states of disease in South Africa.

This chapter examines the ways in which the political environmental context has produced vulnerabilities to the HIV/AIDS epidemic in South Africa. The country has been living with HIV/AIDS for decades, but as I will describe, these experiences are distinctively South African. The political environmental context shapes how particular individuals are made vulnerable to infection while demonstrating how the biophysical setting shapes the opportunities for managing health with improved access to ARVs. Scholarship on the ecological dimensions of human health has been influential in showing how variations in biophysical processes contribute to variations in human health. Yet it is only through the integration of these perspectives that the intersecting social and ecological dimensions of human health can be appreciated. These points of intersection create the conditions whereby some become infected with HIV while others do not. They are points of contact through which HIV is perceived, treated, and managed in rural South Africa. The possibility of chronic HIV is changing lived experiences with the disease, yet HIV lifeways are about far more than accessing ARVs.

CHRONIC HIV IN THE ERA OF ARVS

Before we consider South Africa's experience with the HIV/AIDS epidemic, it is necessary to begin with a basic understanding of the disease.

The human immunodeficiency virus (HIV) is spread through bodily fluids, primarily through sexual activity, shared needles, and other means that facilitate fluid exchange. HIV can develop in the host body over time and lead to acquired immunodeficiency syndrome (AIDS). HIV damages a person's body by destroying CD4+ T cells, which are one measure of immune-system health, and so this count (hereafter "CD4 count") is an important metric for identifying not only the presence of HIV, but also the point at which ART takes place. Individuals infected with the virus can take months or years to become symptomatic, at which time there is a transition from asymptomatic HIV infection to HIV disease. At this early stage of HIV infection, the person does not have signs or symptoms of AIDS such as opportunistic infections, but they can have persistent sweating or fatigue, skin disorders, weight loss, and joint pain. In the United States, the CDC (Centers for Disease Control and Prevention) identifies fairly discrete stages in the progression of HIV from acute infection to clinical latency and finally to AIDS.[8] "AIDS" refers to a group of symptoms that indicate or characterize a disease. In the case of AIDS, this can include the development of certain infections and/or cancers, as well as a decrease in the number of CD4+ T cells that are crucial in helping the body fight disease. This progressive breakdown of the immune system results in opportunistic infections and cancers in the body, thereby causing death. While it remains common to couple HIV with AIDS in discussing the epidemic in various settings, there is a growing conviction that the disease should be considered a *chronic condition* in some areas of the world.[9] This means that access to effective health care and ARVs extends the lifeways of those infected with HIV, thereby preventing the transition to AIDS.

Another key dimension of HIV/AIDS is that the progression from asymptomatic HIV to AIDS is not uniform or direct. How quickly an individual moves from asymptomatic HIV to HIV disease depends on their overall health and can vary temporally. The embodiment of the virus is therefore differentially experienced. Similarly, being HIV-positive is not a fixed state. An individual's health, as measured by the narrow metric of the CD4 count, can vary with factors such as nutrition or the presence of another illness such as tuberculosis. Reductions in the CD4 count can generate severe symptoms that often result in the person getting tested and being put on ART. This has been referred to as the "Lazarus effect," whereby a person sick from HIV has been seen to recover, seemingly from the grave, because of the effectiveness of ARVs. This likely accounts for the way many of my informants discussed the presence of the virus, noting that one of their neighbors "got sick" and then visited the clinic and "got better." Yet other residents have HIV and are not on ART because of their high CD4 counts. In one interview I conducted in 2014, a woman named Mafuane shared that her CD4 count was 650 and so she was taking boosters to help support her immune system. Time and again, HIV-positive individuals turned to CD4 in discussing their health and how they were managing HIV. For them, these counts are a marker of their embodiment of HIV and represent the variable states of disease.

It is unlikely that any country has received more critical attention to the HIV/AIDS epidemic than South Africa. Years of resistance to a national treatment program during the initial stages of the epidemic, coupled with conflicting public health messages about the severity of HIV/AIDS by national officials, resulted in international criticism about the handling of the crisis. It is estimated that currently 6.8 million of South Africa's 53 million citizens are infected with HIV, which means that there are more people living with the disease than in any other country in the world.[10] The Joint United Nations Programme on HIV/AIDS (UNAIDS) reports that roughly 18.9 percent of South African adults between the ages of fifteen and forty-nine are believed to be HIV-positive, of which 3.9 million are women and 340,000 are children under the age of fifteen. As a result of the HIV/AIDS epidemic, there are an estimated 2.3 million orphans under the age of eighteen.[11] These estimates reflect many factors, including the rapidity of the spread of the disease and the mishandling of the epidemic by the South African state during its early stages. The first reported case of HIV in South Africa was in 1982, and from that point forward the country experienced a surge of new cases, cresting with an estimated infection rate of 18 percent in the early 2000s. It must be noted that these "national" estimates belie the social and spatial variations within the country, thus obscuring deeply rooted dynamics that shape the landscapes of HIV.[12] These official estimates were also a part of the political terrain during the era of President Mbeki, who stubbornly held to previous statistics from the United Nations rather than the updated ones. Some scholars at this time suggested that HIV/AIDS discourse was tied to the process of decolonization and post-apartheid transition, whereas some government officials asserted that the disease was presented in the discourse of Western donors and corporations to depict Africans as immoral, sexualized, and racialized.[13] These depictions were believed to be designed to sell pharmaceuticals—which was not easily countered, given the resistance of major medical corporations to releasing generic ARVs, irrespective of international patent law. Historically, paranoia about disease outbreaks was manipulated by colonial and apartheid regimes to justify racial segregation.[14] As a result, there is understandable skepticism about the motives for certain public health interventions.

As the epidemic expanded in the 2000s, pressure on the national government mounted to more publicly address the disease and provide access to drugs, as other African governments had done.[15] By this time Uganda had become an international leader for its "ABC program" that emphasized abstinence, being faithful, and using condoms or other preventative measures. In 2000 in South Africa, AIDS accounted for 25 percent of all deaths, and mortality was 3.5 times higher than in 1985 among women twenty-five to twenty-nine years old, and two times higher among men thirty to thirty-nine years old.[16] National data sets showed that total deaths (from all causes) increased by 87 percent between 1997 and 2005, with death rates tripling for women aged twenty to thirty-nine, and more than doubling for males aged thirty to forty-four. At least 40 percent of deaths were believed to be attributable to HIV/AIDS. The delay in aggressively responding to the epidemic left a tragic toll for the country's population and continues to inform the legacies of past governmental officials. A report from the Harvard School of Public Health in 2008 faulted the national government for its delay in distributing AIDS medications and estimated that at least 330,000 people had died and tens of thousands of children had been born with HIV because of the denial of access to nevirapine.[17] At that time, becoming infected with HIV in South Africa was a death sentence, while elsewhere it could be treated through the use of ARVs.

The political transition to President Zuma's administration was accompanied by statements less focused on contesting estimates, but encouraging testing to determine one's status and becoming aware of treatment options. In a speech delivered before Parliament in 2009, Zuma stated that all South Africans must know that they are at risk and must take informed decisions to reduce their vulnerability to infection, or if infected, to slow the advance of the disease. Most importantly, all South Africans need to know their HIV status, and be informed of the treatment options available to them.[18]

A marked shift in recent years has been increased access to ARVs that, even with hundreds of thousands of new infections per year, are being provided to almost two million people each day. The effect of this expanded national commitment to care is that the national life expectancy has increased by eight years since the mid-2000s.[19] The 2012 UNAIDS country-level progress report stated that the epidemic has reached a "plateau" in the country, with HIV prevalence for adults aged fifteen to forty-nine years being 15.6 percent in 2002, 16.2 percent in 2005, and 16.9 percent in 2008. According to data on antenatal care, HIV prevalence had "gradually leveled off" below 30 percent after increasing from 7.6 percent in 1994 to 29.5 percent in 2004. While the rate of HIV prevalence had flattened, the absolute number of people living with HIV was "on a steep increase" of approximately 100,000 additional people each year. Finally, the progress report noted a "substantial downturn in AIDS related mortality in recent years," with the annual number of AIDS deaths reduced from roughly 257,000 in 2005 to 194,000 in 2010.[20]

Integral to the provision of these drugs has been a decentralization of care down to local health facilities, including rural clinics, in providing ARVs and other services. Clinics and home-based care groups are on the front line of testing and surveilling communities to ensure compliance with treatment protocols. The types of drugs have become easier to take and the government has recently advocated a course of treatment that includes one pill taken once per day for only hundreds of dollars per patient per year.[21] Although I began the chapter by discussing the challenges to universalizing access, it is clear that the South African government has made remarkable progress in recognizing the severity of the epidemic and ensuring access to ARVs throughout the country. While impressive in extending the lives of infected people for years and decades, the expansion of ART presents other challenges for HIV management, including new resource needs and shifts in the ways that people understand their individual and community health.

There is not a consensus within the scholarly literature that improvements in accessibility of ARVs mean that developing countries have reached a new stage of chronic disease management. In reviewing the concept of HIV chronicity, Janet McGrath and colleagues suggested that the paradigm of chronicity emphasizes self-care, biomedical disease management, social normalization, destigmatization, and uncertainty. In advocating a new paradigm for HIV chronicity, they showed how HIV differs from other chronic diseases in resource-poor settings "because the economic burden of treatment compounds daily struggles and treatment regimens challenge people's ability to maintain privacy and manage stigma."[22] Chronic HIV has unique features compared to other chronic conditions, particularly within developing regions where greater access to ARVs has only recently arrived at levels where it is even possible to talk about management of the virus. McGrath and colleagues noted that HIV differs from other chronic diseases in that HIV remains infectious within the individual body, it retains persistent social stigmas, and HIV-infected individuals experience multiple episodes of other illnesses and opportunistic infections. In a related assessment, Carl Kendall and Zelee Hill emphasized that "HIV/AIDS continues to be experienced by many, through intermittent acute episodes of symptoms of disease, with underlying worry about disease progression and worry about the perpetuity and life-threatening nature of the illness, rather than a relatively low-level, long-lived condition 'under control.'"[23]

HIV lifeways vary at the bodily level and are produced by underlying social conditions that challenge the possibilities for normalization and effective disease management. The states of disease mean that individual experiences are dynamic, such that the domain of health can shift over time and space. Because of this, throughout the book I will focus on what I call *managed HIV* to describe the lived experiences of those navigating the public health infrastructure for treatment and access to other resources critical to survival. Managed HIV is not the same as chronic HIV. Regardless of the state's assertions of widespread access to ARVs, it is not the reality for all of its citizens. In South Africa as elsewhere, people lack access to treatment regimes that offer the possibility for survival. Access is mediated not only by one's spatial location and proximity to clinics, hospitals, and other sites of care but also by socioeconomic class and gender. While public health institutions advocate universal access to ARVs, stigmas are not equally experienced. Men and women have different anxieties about the virus, which shape their approaches to testing and their adherence to ART. The resulting discussion is meant to challenge the linking of HIV chronicity with declarations of universal access, particularly due to the social and ecological dimensions of disease vulnerabilities and constraints in health services that produce irregular, uncertain, and inequitable HIV lifeways.

MANAGED HIV IN SOUTH AFRICA

My engagement with managed HIV is centered in Mpumalanga Province in northeast South Africa. Mpumalanga is situated in the eastern *lowveld* (lowlands) of South Africa, extending south of the Kruger National Park, west of Mozambique, and north of Swaziland (see figure 3).

Urban and peri-urban areas include the provincial capital of Mbombela (formerly Nelspruit), Malalane, and KaNyamazane. The region has been an important national agricultural center and, as such, has been implicated with processes of racial segregation and dislocation for decades. During the late 1940s and '50s, territory in the fertile catchments of the Komati and Crocodile rivers was removed from African control and given to white farmers to produce citrus and subtropical crops for regional and international markets.[24] Sugarcane production was more widely practiced from the 1970s onward, and for several decades this crop was grown in conjunction with fruits and vegetables for foreign export. During the apartheid era, agricultural production was dependent on labor from Mozambique and the neighboring Bantustans, particularly KaNgwane, which encompassed most of the Nkomazi administrative district.[25] Following the 1994 democratic elections, the national government began promoting sugarcane production for what it identified as an "emergent" African farming class. This has meant that territory from Malalane to the Swaziland border has been converted to small-scale farms, in some cases seven hectares or less, and also larger-scale farms, much on privately held land outside the borders of the former KaNgwane Bantustan. One such sugarcane project was initiated in the Mzinti community in 2000 as part of the Land Redistribution for Agricultural Development Programme

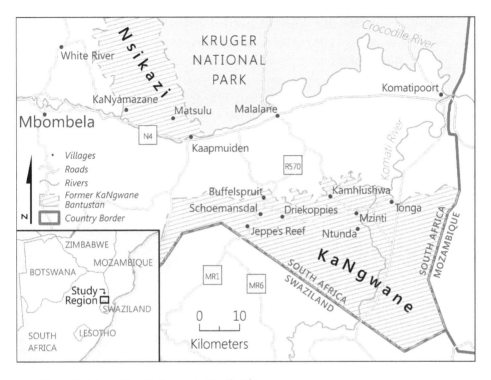

Figure 3. Map of the study area in South Africa.

(LRAD). LRAD was promoted by the Department of Land Affairs and the Department of Agriculture, Conservation, and Environment with the purpose of generating irrigation development for 1,828 hectares of farmland for the settlement of 241 small-scale commercial sugar farmers, 456 women's group members, and 50 youth-club members. An area of roughly 245 hectares north of the Mzinti community was converted to sugarcane farms in 2001. This area was divided into thirty-five plots of seven hectares each, with a variety of grants and loans forming the basis of the LRAD plan.[26]

I have concentrated my research within three villages in the area that serve as case studies for understanding the relationships between health and environment, with a particular focus on HIV/AIDS. As shown in figure 3, the case-study villages are Mzinti, Ntunda, and Schoemansdal. Although they are located in rather close proximity to each other, there are differences between the three villages that facilitate a comparative analysis of the impacts of HIV/AIDS on social and ecological systems. First, the villages differ considerably in population size. Ntunda is the smallest of the three and is considered "more rural" by residents because of its location farther south of a main road that runs from One Tree Hill to Tonga. For most of the fieldwork period, Ntunda did not have a clinic. This meant that residents had to travel five kilometers north to Mzinti, or in some cases farther afield to other villages, to access clinical medical care. After delays, the Ntunda clinic opened in 2015. Schoemansdal is the largest of the three case studies and is a bustling political and economic center in the region. Its location on route 570, which runs from Malalane to the Jeppe's Reef border gate with Swaziland, ensures consistent commercial and tourist traffic through the area. The Matsamo Tribal Authority is headquartered in Schoemansdal, giving it political heft compared to other villages in the region. Since it is also the site for distributing social services such as pensions and child grants, Schoemansdal is an economic hub in the area.

In terms of population size and economic activity, the Mzinti community is situated between Ntunda and Schoemansdal. Additionally, including Mzinti as one of the case studies had the advantage of allowing me to integrate long-term fieldwork to provide a longitudinal perspective on health–environment interactions.[27] Since the time of my earlier research, the village has expanded, and there are more shops and commercial activity in neighboring Tonga just across the river to the east. The three communities are spread throughout the region and are differentially affected by patterns of migration to peri-urban centers. Mzinti and Schoemansdal are adjacent to roads that provide public transportation to Malalane, while Ntunda is more remote, so households in this village are more likely to have different livelihood systems because of the distance from employment centers. While all three communities fall under the jurisdiction of the Matsamo Tribal Authority and seemingly follow similar mandates in terms of natural resource collection, the three communities are proximate to different vegetation types. Mzinti and Ntunda are situated in *lowveld bushveld* vegetation, while Schoemansdal is located in *lowveld sour bushveld*.[28] Lastly, the three communities are differentially positioned in relation to water sources for agricultural production, particularly in terms of the Komati and Lomati rivers that support dryland farming and sugar-cane production.

The South Africa UNAIDS progress report specifies the drivers of the HIV/AIDS epidemic to be intergenerational sex, multiple concurrent partners, low condom use, excessive use of alcohol, and low rates of male circumcision.[29] Based on my interviews with health-care providers and NGO workers conducted in 2006, there are additional factors believed to influence the rate of infection in the area. First, a lack of employment opportunities contributes to migration into Mozambique and Swaziland and nearby peri-urban and urban centers. This is believed to increase HIV exposure in rural areas, particularly along the borders. Second, there remain persistent stigmas surrounding condom use that, coupled with limited sexual decision-making power for women, restrict the ability to reduce transmission rates. Third, the health-care infrastructure is less developed than in other regions of the country, which is partly a product of historical development planning. This area suffered from comparatively limited governmental investment in infrastructure and social-service delivery during apartheid, and this continues to influence the availability of health-care services. The consequence is that clinics and home-based care organizations continue to play critical roles in the distribution of information and the provision of social services. This has changed somewhat in recent years as the national government has taken a more active role in addressing the HIV/AIDS epidemic, but the role of nonstate organizations remains essential in providing services to rural residents.

My objective in this chapter is not to interrogate the history of HIV/AIDS in South Africa during a particular period or as lived experience under one governmental administration. While the tumultuous period of 1998–2006 is notable in understanding the spread of HIV and how the epidemic has been differentially experienced within the country, this history has been the subject of considerable attention elsewhere.[30] Rather, I want to examine the ways in which the political environmental context produces vulnerabilities to infectious disease and mediates the possibilities for HIV management in the contemporary era. In chronicling the effects of increased access to ART, I seek to uncover the lived experiences of individuals, families, and communities in responding to HIV/AIDS, as well as their perceptions of health and well-being more generally. These

HIV lifeways now exist within a national pharmacological infrastructure that has become more aggressive in providing ARVs. As is happening in other parts of Africa, ART is turning HIV into a managed condition for many of those infected, which means that having HIV is not necessarily a death sentence. Yet access to ARVs is shaped by underlying structural conditions, and their use is conditioned by other social and environmental needs not equally experienced. The health domain varies in ways that are linked to broader structural conditions and spatial economies that have been produced over centuries in ways that continue to mediate the lifeways and landscapes of managed HIV.

"HIV IS NOT AFFECTING PEOPLE— PEOPLE ARE AFFECTING HIV"

Upon returning to the area, it was necessary to begin by learning the ways in which health and well-being are understood within these communities. The tumultuous period associated with the presidency of Thabo Mbeki generated particular views of disease and efficacy of treatment options, but it was possible that these understandings had shifted in recent years. While expansive, the HIV/AIDS epidemic is not static. Its features move with competing understandings, new interventions by political institutions, technological and scientific discoveries associated with the disease, and fluid social relationships that facilitate the flow of knowledge, capital, and exchanges across networks. Vulnerabilities to disease and the possibilities for healthy decision making are similarly dynamic and are the product of negotiations between social actors, institutions, and the various structures in which they are situated. In previous years, HIV/AIDS had been a disease hidden in plain sight. Now it was important to question whether HIV had become more visible and in what ways it is now understood, embodied, and managed.

This and the two following chapters are informed by research conducted over fifteen years, utilizing a mix of qualitative and quantitative methods. My first experience in the region was in 2000 as part of a preliminary trip in advance of my doctoral fieldwork that began in earnest in August 2001. This work was centered in the Mzinti community, where over the course of twelve months I conducted qualitative interviews with family members and conservation organizations, participant observation of livelihood and resource-collection strategies, and a structured survey of 478 households that was administered by a team of research assistants with whom I continue to collaborate. As discussed in the Introduction, the objective of my research at that time was to examine livelihood and environment dynamics in the former Bantustan territories in order to understand the ways in which historical spatial economies were being transformed following the 1994 democratic elections. Subsequent visits to the region occurred in 2004 and 2006 and included interviews with healthcare providers and representatives of NGOs and visits to clinics and home-based care organizations. This information is supplemented with findings from November 2012 to June 2016 that were generated from participant observation, qualitative interviewing, household surveys, and focus groups conducted either by me or by members of my research team. Structured surveys were conducted with 327 randomly selected households from the three villages, gathering demographic, education, economic, and environmental information on 1,546 individuals. Qualitative interviewing has been conducted with respondents who participated in the structured survey, and I draw upon some of this information in discussing the social and ecological dimensions of managed HIV. The combination of methods allows for a nuanced portrait of the lived experiences of residents in the study area and the shifting views on HIV/AIDS in the contemporary period. These methods also help reveal critical social and ecological dynamics that produce the political environmental context, which continues to generate disparities in vulnerability to disease and the possibilities for managing health in the era of ARVs.

In order to address current discourses of health and well-being, six focus-group interviews were conducted in the three villages of Mzinti, Ntunda, and Schoemansdal in 2012. The focus groups were supervised by Dr. Margaret Winchester, who at the time was a postdoctoral research scholar at the Pennsylvania State University, and my research assistants Thami and Elinah.[31] In each of the villages, one focus group was conducted with men and another with women. The reasoning for this is that it can reduce tensions between men and women that might exist in the setting, thereby supporting more open discussion. It also recognizes that

men and women often have unique, and sometimes competing, understandings of disease and health based on the different roles they play in
their families and communities. Because labor roles can be highly gendered, the differential social and economic responses to health challenges
are also likely to be variably understood. Men and women feel the impacts
of disease in varied ways and embody it differently both materially and
symbolically; in essence, their lived experiences of disease and health differ. Participants were selected to ensure diversity by age, level of interaction with home-based care groups, employment, socioeconomic class, and
residence history. The number of participants varied by focus group, with
the smallest having eight participants and the largest having fifteen.
Altogether, the focus groups included sixty-seven participants from the
villages of Mzinti, Ntunda, and Schoemansdal. All of the focus-group
interviews were tape-recorded with the permission of participants, and
full transcripts were created through notes taken by Thami and Elinah
and in conjunction with the audio recording.[32]

The focus groups began with the facilitators, Thami and Elinah, asking
the participants to list the most pressing health concerns for the community, with the goal of ranking them in order of their perceived importance.
The facilitators were instructed to not introduce HIV into the focus
groups, in order to see if, and when, participants would raise the subject
themselves. In fact, one of the six focus groups, a group of men from the
Mzinti village, did not identify HIV as a specific health challenge for the
community. In this case, Thami and Elinah were asked to raise the topic
to ensure that it was addressed by the group. After the listing of pressing
health concerns, the facilitators asked the group to discuss them, high-
lighting the ways they felt these health problems were spread, which people were affected, how individuals should respond, and the types of
resources available. Following these questions, the focus groups were
asked about HIV in particular, with an emphasis on HIV in the community. Participants discussed how HIV was spread and could be prevented,
the types of effects it had on individuals and communities, who was
affected the most, and the types of resources available to help those with
HIV. The final component of the focus groups was to investigate the possible relationships between human health and the natural environment,
to examine the intersections between social and environmental impacts of

disease. The following sections detail some major themes that arose from the focus groups, structured surveys, and my qualitative and ethnographic fieldwork in the region that began in 2000. The themes speak to how the states of disease are producing social vulnerabilities and health possibilities. The emphasis is particularly on HIV/AIDS, but as the discussion demonstrates, while the medical and policy communities focus on biomedical HIV, people infected and affected perceive and respond to the virus in more complicated ways.

"PEOPLE ARE SICK"

Within these communities, people have HIV but they do not disclose their status, at least not readily or willingly, especially to strangers or to those outside their trusted social network. But understandings of health are foregrounded, because time and again I am told that people in the community are "sick." Illness is present and is the language used to discuss HIV. My first lengthy exchange with Nelson, my long-time friend and collaborator, went along these lines. I have known Nelson since my doctoral research in Mzinti in 2001, and he has worked as a close research assistant and collaborator since that time. Upon returning in 2012, I explained that I was intent on uncovering the relationships between health and the natural environment. I shared that I wanted to speak with people in the community who were HIV-positive and who were taking ARVs. He nodded his head solemnly but did not speak of HIV. Nelson indicated that there were people in the community who were "very sick" and that in some cases they "got better." Presumably this referred to their being tested for the virus, most likely at the local clinic, and then receiving ART. I was struck by his reluctance to say "HIV"—he preferred instead to talk about "sickness." We were joined later in the day by Thami and Elinah to informally discuss health in the community. Thami displayed a similar note of caution in mentioning HIV, preferring to say that people were sick. He also emphasized that if we were interested in health, then we should talk with traditional healers because they would be the ones to answer our questions. This suggestion did not surprise me, because I knew that Thami had extensive knowledge of traditional medicine for someone of his age.[33]

When the structured surveys were conducted in Mzinti in 2002, Thami would return with forms that had impressive detail about the reliance on specific types of traditional medicine, outlining what was used and for what purpose. A generally quiet person but an exceptionally keen observer, Thami was effective in getting respondents to speak about the different types of care they would pursue, including visiting traditional healers or utilizing traditional medicine. Thami took lengthy notes on the range of medicinal plants and other courses of treatment on which respondents depended, and he was intent on sharing this with the research team to inform what we were learning about natural resource collection in the community.

In contrast to Nelson and Thami, Elinah was more vocal in discussing health and was the first of them to openly say "HIV." This was likely the result of her unique positionality, shaped by her work with a clinic in Schoemansdal for several years. I have known Elinah since 2001, and over that time have been amazed by her unmatched knowledge of community dynamics and histories. I had last seen her in 2006, when she was volunteering with a youth group that was working to disseminate information in the area about HIV. At that time, the clinic had an active home-based care group that was providing services to the sick and dying in Schoemansdal and other villages. Schoemansdal is one of the largest villages in the area, and it was common to see provincial governmental vehicles, NGOs, and development groups moving throughout the area. At the time of my visit in 2006, there were several offices on the grounds that were bustling with activity. One office contained a carpenter who was building coffins that ranged from adult size to one for a child no older than six. During my tour of the facility, Elinah opened a second office that contained shelves of medications that the home-based care group would distribute to its clients. I was impressed with the amount of materials that were on hand because it suggested a reasonable flow of external capital to support the activities of the organization. However, I was also struck by the reliance on young members of these communities to visit with patients and distribute medications. It suggested a decentralization of care, whereby external funders and the national government were relying on local residents to provide critical health services to rural populations. In her work with the youth group, Elinah proudly explained that she had led educational workshops about health and that she understood the ways in which the epidemic was affecting individuals and their families. She noted that she had also worked in Swaziland on a public health campaign and would be comfortable facilitating the focus-group discussions.

When people discuss illness, they are commenting on their individual health but also their position within broader structures of power and domains of meaning. The statement "He was sick" speaks volumes about how health is understood and the ways in which care is pursued. The health domain is not simply material, referring to the types of services that a person might be able to access. It is also symbolic in how it communicates a person's status, how a community is perceived, which social actors are deemed culpable for spreading a virus, and the familial and social networks in which a person is situated. Often when residents talk about their individual health, they are also referring to their family, to their social network, and to their community. These extensions of the self provide a sense of well-being and help offset the costs incurred by attending to disease or other challenges. Other work from South Africa has shown that HIV is interpreted by some residents as a disease of the community, in that it reflects a general decay of shared well-being. These disease perceptions, which are localized and contextual, can be interpreted in cultural terms such that HIV transmission is understood as an erosion of cultural norms and traditions.[34] In order to capture these dynamics, one strategy employed in the focus-group interviews was to ask people questions about the community rather than directly about themselves. Participants were free to share what they wanted, and it was clearly communicated that all identities would be kept anonymous, but for the most part there was greater comfort in talking about health in more general terms. This had the effect of directing the discussion to the community level, so that disease was seen as a community problem in addition to an individual one. People were more willing to talk about HIV as affecting their neighbors and their communities, whereas the intimacy of the disease, coupled with past stigmatization of HIV-positive individuals, made it a challenge to speak openly of the virus in personal terms. Thus, there was an "othering" of the virus. It should be noted that this was not universal, and there were many who spoke openly of their status. But generally (and understandably), there was a privacy around the topic and a reluctance to disclose individual status to others, perhaps for fear of the consequences, or simply a desire to keep it to oneself.[35]

Research completed during the initial years of the HIV/AIDS epidemic in South Africa found that people had a diversity of opinions about what caused the disease, ranging from transmission through physical contact to its being a traditional sickness caused by the erosion of cultural mores, to conspiracy theories that it was spread by white doctors to infect Africans, to a belief that it was the result of witchcraft.[36] In the six focus groups, respondents consistently talked about the disease in biomedical terms as being spread through HIV. While there was an intermingling of the disease with traditional practices and community well-being, understandings of transmission were narrower. One woman explained:

> I am not afraid of this condition and also not afraid to live with the person infected. Just the only thing is to take care of yourself and use condoms every time, to play safe. HIV is something everyone talks about. The government, especially the Department of Health, is always having talk shows and HIV and AIDS is mentioned in the radios, also on television and the press news.

Another woman from Ntunda explained that the village has knowledge about the disease because of the home-based care center that provides help and information. She explained that "HIV is here in Ntunda. We know it, we have knowledge about it now. But in the beginning we did not have knowledge. Again, since the home-based caregivers are present, they also help us and give us information about how HIV infects us."

The smallest focus group was conducted with eight men from the Mzinti village, most of whom were in their twenties. This was the only group that did not mention HIV when listing health concerns for the community, but after being prompted they were willing to discuss the subject. By comparison, they suggested that diabetes and tuberculosis were the most significant diseases in the community. Members of this group explained that there was no longer a stigma surrounding HIV because people knew more about it now and often had to care for sick relatives. One person noted that the youth were more likely to be affected because they had multiple partners and might lack education about using condoms or be under the influence of alcohol. It was emphasized that young men might compete with each other to have the largest number of sexual partners. Regardless, one man mentioned that HIV could be prevented by using condoms, having one partner, being open about one's status, and education. The group also explained that HIV could be treated with drugs and immune boosters taken with proper food. What is notable is that HIV transmission was presented as preventable as long as appropriate measures were taken, such as being faithful to one's partner and using prophylaxis. These statements suggest the acceptance of a behavioral perspective on HIV. The epidemic is centered at the individual body, thereby emphasizing the decisions taken by each social actor in negotiating sexual decision making and intimacy. Should the virus be acquired, it is treatable using drugs available through the public health infrastructure. I return to the biomedical HIV perspective in a later chapter because it is present in the communities, particularly given the increasing access to ART. Yet acquiring and using ARVs is not a simple matter. These lifeways are secured by navigating a complex landscape that shapes the possibilities for survival.

Earlier I discussed the structured household survey that was conducted in the Mzinti community in 2002. In that survey, only one household out of 478 surveyed attributed the death of a family member to AIDS, yet there was a measurable reportage of adult mortality due to tuberculosis, which was likely associated with the weakened immune system of infected people. By contrast, participants in the 2013 survey were more willing to disclose their HIV status. In the first completed survey in the village of Schoemansdal, a woman roughly forty years of age shared that she was HIV-positive and was currently taking ARVs. She was willing to talk about her general health with Mabiso, the team member who administered the survey, sharing that she worked at a nearby health facility and that this position had provided information to her about seeking services for HIV. She thanked us for the work that we were doing and was markedly optimistic about her health situation, expressing her desire to see us on our next visit to the community. Of the 327 households surveyed, the percentage of interviewed heads that indicated they were HIV-positive was 15 percent. This is lower than the national estimate of 18 percent but represents a marked contrast to the 2002 survey, when there was a greater reluctance to openly discuss HIV/AIDS.[37]

Nearly 90 percent of respondents who shared their positive HIV status indicated that they were on ARVs. This implies a number of things. For one, it suggests that the country's efforts to universalize access have had some success. While there remain challenges in receiving consistent access to ARVs at nearby facilities, the availability of these drugs is changing the landscapes of HIV through which people navigate in seeking out care and needed resources for survival. Second, greater access to ARVs is changing the ways in which people understand the disease and likely shifts their level of comfort in discussing their individual health status. As the survey was being conducted, I had conversations with members of the team about stigmas surrounding HIV and the willingness of community members to be open about their status. Nelson explained that stigmas about the disease remain, but that people should be more open: "Psychologists are saying that people should be open about HIV because if they hide it can make things worse. You might think that people are watching your every step." But he explained that if you are open about it you can get counseling, and with the proper drugs you can live for years. He said they might point to someone who has been living for fifty years with HIV and say, "Look at that person. She is on ARVs and she is fresh." This coupling of HIV with ARVs is a discursive shift from the time of Mbeki's administration. Nelson and others throughout the villages invoke the specter of the virus in new ways that include a biomedical framing whereby "sickness" is replaced by improved health status made possible by ART.

As our conversation continued, I was reminded of a comment made by another member of the survey team during our three-day training workshop. We had begun talking through the survey questionnaire and had come to the questions about HIV status and use of ARVs. The team was consistent in noting that these were challenging questions and that people might have difficulty answering them. One member, Mabiso, had previously worked on public health campaigns in Swaziland attending to HIV. He was also a youth minister in his village and so had unique experiences of interacting with others in the community. Mabiso noted that people in Schoemansdal were aware of the disease and that they would be willing to speak openly about it, though not as much as those in other parts of the country, such as in the province of KwaZulu-Natal. Mabiso said that people there say they are more afraid of a dangerous neighbor than of HIV because the dangerous neighbor could come along and kill everyone. By comparison, "HIV can take years to kill someone." I was struck by how these conversations indicated that the discourse of HIV was shifting. Time and again, people talked about HIV and treatment in the same terms, combining them in ways that seemingly altered understandings about the disease. People talked about living with the disease and identified it as a longer-term challenge. With the distribution of ARVs in the rural areas, the epidemic has become more visible in the lived experiences of residents and within the landscapes they inhabit.

Another factor contributing to shifting discourses around HIV are perceptions of the disease within the community. As has been reported in other African countries, it is likely that if community members think others have the disease, this could have a leveling effect in making it easier for people to talk about their own status. Survey respondents were asked how many people in their community did they think had HIV, and more than 60 percent said "most people/nearly everyone" and "very many." Another 20 percent estimated that about half had HIV. Thus, 80 percent of the 327 respondents believed that half or more of the community had HIV. When asked about ARVs in the community, more than 70 percent reported that about half of the community or more were using these drugs. This is a striking paradox: while less than 20 percent of household heads reported their HIV status as positive, they believed that more than half of the community had the disease. This suggests a normalization of HIV in the communities, where it is seen as part of a range of other health and social interactions. The landscapes of HIV are thus interpreted in complex ways, based on perceptions of individual and community well-being that are interlinked with other dynamics necessary for maintaining HIV lifeways.

"WE ARE EATING THEIR FOOD": CARE AS MULTIFACETED INTERVENTION

Seeking and providing care for HIV-infected people is not simply a matter of accessing institutional systems or critical medical resources. Given the complexities of managed HIV, care includes considerations such as nutrition and caloric needs, food security, and governmental documentation. These are

among many factors shaping individual experiences that require engagement with the political environmental context of managed HIV. While
increased access to ART has saved lives in South Africa, it has also presented
new challenges that are structured by the existing social and ecological systems in the region. These points of contact make possible the securing of
life-saving drugs and other forms of care but also the possibilities for a particular quality of life. With the expansion of ART in Sub-Saharan Africa,
emerging research is documenting the experiences of those receiving care. In
groundbreaking work from Mozambique, Ippolytos Kalofonos found that
the scaling up of treatment collided with preexisting social conditions associated with historical dynamics and socioeconomic poverty.[38] He pointed out
that hunger is the most commonly cited complaint of those on ARVs, which
is related to the increased nutritional demands associated with treatment in
addition to previously unmet needs. He asserted that this serves as an
"embodied critique," a reminder that "even while they save lives, AIDS treatment programs can paradoxically have dehumanizing effects if the broader
social structures that contribute to suffering and impoverishment remain
hidden and intact."[39] Although ART has been coupled with nutritional education that is provided by community associations and clinics, Kalofonos's
research suggests that this is not enough to meet the demand.

Similarly, messages sent from clinics and hospitals in the area emphasize the need to access certain foodstuffs to maintain good health, not all
of which are readily available in rural areas. HIV-positive individuals are
encouraged to plant gardens to acquire healthy foods, but they are warned
against mixing traditional medicine with ARVs. According to the Food
and Agriculture Organization, "food security exists when all people, at all
times, have physical and economic access to sufficient and nutritious food
that meets their dietary needs and food preferences for an active and
healthy life."[40] While increasing awareness of nutrition is a positive outcome for affected families, food security is produced by social and ecological conditions that have been constructed in particular and contingent
ways. Managed HIV is achievable only by successfully navigating the
landscapes of care. One female participant in the Schoemansdal focus
group, who was a home-based care volunteer, raised concerns about the
relationships between socioeconomic poverty, food insecurity, and nutritional needs. As she explained:

I have lived here since nineteen ninety-one. Like my village, I'm starving, living in poverty. I am volunteering here in this home-based care. I thought maybe when time goes on we will get some assistance and I continued to volunteer. People come and go just like you, as you did today. People promise things but they don't fulfill them. Nothing is happening. These things that we see are not good, it is so painful. When we visit homes you can also feel the pain. Sometimes when we go to the patients to give them food like spinach, since we have this small garden, they will call us and say that we are eating their food. Sometimes you will be afraid to enter in some of the patients' homes because you don't have food and you know that they are taking pills. Because we sometimes try to give them soft porridge but we need more and we have no money.

Another female participant echoed these concerns and shared that patients were unemployed and had no food to eat. She emphasized the importance of proper food, like instant porridge, fruits, and vegetables, for those attending to health needs in the community. "Getting better," as some residents explained, requires accessing ARVs, but it also requires other resources to ensure a healthy quality of life.

I conducted an interview with Mafuane, in her Mzinti RDP home, that emphasized the entanglements of HIV status and food security.[41] In visiting the clinic every six months to assess her CD4 count, she was routinely encouraged to eat healthy and nutritious foods. In the event that these were not available, like others in the community, Mafuane was encouraged to grow a garden. We concluded our conversation by discussing some of the continued challenges for those who are HIV-positive. When asked about the role of the government, she commented that the pills were being provided, and those who were sicker and unable to find work received food parcels. Mafuane emphasized that the government was helping, but that it must provide jobs or food parcels to more people because a lot of HIV-positive people were unemployed. This resonates with the statements of many with whom I have spoken: It is not enough to provide ARVs. The government must provide jobs. There must be food. ARVs without food are not enough.

Navigating the health-care landscape requires certain forms of status and documentation that facilitate encounters between the state and its citizens. It is common for residents to note that accessing health care was constrained by not having the proper governmental identification. This is not surprising, given the proximity of the Nkomazi district to the Mozambique and Swaziland borders and the amount of migration between these countries. For example, a member of my research team described the Mozambican section of Mzinti, which he believed was comprised of residents who fled the civil war and likely crossed the border illegally. When I was interviewing families in 2001 and 2002, participants would sometimes indicate that their families fled the civil war and had settled in Mzinti. In speaking with them, it was clear that they were concerned about xenophobic responses from others in the community and were therefore careful about sharing their family history. By comparison, the village of Schoemansdal is roughly ten kilometers from the Jeppe's Reef border post to Swaziland, and the unfenced border contains several crossing points for dispersed family networks to visit with each other. As such, some residents in the area lack the official government documentation necessary for accessing social services and economic resources. As one woman from the Mzinti focus group explained, without the proper identification it is difficult to get services. She stated:

> There is a granny from South Africa who got married in Swaziland but because of family problems she came back to South Africa. The granny is staying with her brother's children. One is twelve years old and the other is fifteen years old. These kids do not have identification and they are schooling. . . . The granny tried all of her means to talk to social workers but it did not work. Meaning that the Department of Social Services is not doing its job. This means poor services. This is something that usually happens here, especially if you come from Maputo or Swaziland.

The Mzinti men's focus group also noted the challenges of government documentation, explaining that many could not get grants and pensions because they did not have a government-issued identification card. They insisted that this caused families to starve or young people to get HIV because they needed money and would sell themselves. The group noted that IDs were needed to apply for housing, which was granted by the provincial government and also the tribal authority.

The male focus group in Ntunda had a vibrant discussion about health and environmental issues in the village. This group had thirteen participants who met in the office of a home-based care group that provided services in Ntunda. Members of the group emphasized that the lack of a clinic at that time was an issue because it resulted in people seeking out care from traditional healers. They noted that HIV cannot be cured by traditional medicine, but for other diseases it was possible to receive assistance. The absence of economic opportunities was emphasized as related to the possibilities for health and well-being in the village. As one participant explained:

> Most of the people here are not working and they need jobs. There are those who need jobs but whose lives are better than others and there are those who don't have anything at all because there are no job opportunities. That ends up causing crime and that is because of hunger. I think if people could get food there would be no crime.

This was echoed by a second member of the group, who commented that most of the homes in the village had children as household heads because the parents had died. He suggested that this contributed to crime because they would break into shops to get food, or become pregnant to obtain government grants. Another member of the group shared that when the government created development projects, the stipends associated with them were too small to survive on. He commented:

> The biggest concern here is that the youth are now engaging in alcohol drinking and that is not right. I am also concerned about our tribal authority. You will see lots and lots of trucks coming in to collect sand, which at the end is causing soil erosion. Also, it causes animal problems in terms of grazing because we no longer have land for our animals.

To understand some of the relationships between health and food production, I spent an afternoon with Jablane, a young woman from Schoemansdal whose family owned a farming plot on the western edge of the community. I had encountered her because her family had reported owning a farm in the 2013 household survey and my intent was to learn more about the ways in which members of the family supplemented their diet. Owning a farm is highly unusual because of the legacy of colonial and apartheid spatial planning that provided little farming land for residents in the densely populated Bantustans.[42] Rather, it is more common for households to have gardens on their plot where a variety of crops are grown, such as corn, onion, peanuts, sweet potatoes, and spinach. Additionally, trees are cultivated to provide fruits such as guava, papaya, and banana. We walked to the farm from the main road running to Jeppe's Reef, weaving our way through narrow dirt roads that were the main pathways to the houses in this part of the community. This section spreads toward Driekoppies Mountain, part of the Drakensberg Mountain Range that runs north and south from Swaziland through Mpumalanga, and I could see smoke in the mountains from fires. When I had observed this previously, I was told that the fires were likely made by residents who were hunting animals for food. Upon arriving at the farm, the challenges the family faced in maintaining the plot became immediately apparent. The fields were located adjacent to other houses and were under pressure for other types of use and development. Grazing cattle were visible, and I was told they occasionally moved onto the plot because other community members removed the fence that circled the farm. Jablane explained that they talked to the neighbors about these problems but were rebuffed. Their situation gave the impression of a green island surrounded by competing interests. It felt tense at this moment as our eyes scanned the area, as though her family was threatened by neighbors living adjacent to the property.

In discussing the farm with Jablane, it became clear how its management was mediated by the health and well-being of the family. Her elderly grandmother, who owned the plot, had been in the Shongwe hospital the previous year, which had directly affected the family's ability to plow and plant crops. Jablane explained that normally the family did not plant if someone was sick because, according to tradition, "we must mourn until he or she is healed." As for plowing the fields, that cost sixty dollars to hire a tractor to plow and prepare the ground for growing pumpkins, beans, and corn.[43] I was also interested in the fact that the family collected insects from the fields, which they used to supplement their diet. Jablane indicated that they collected these insects as many as four times in a year, early in the morning, estimating that roughly one hundred were collected at a time, to be fried and then eaten by the family. This pattern has been reported in other parts of Mpumalanga Province, and my visit with Jablane suggested that the hunting of wild game, fishing, and insect collection are important for supporting food security for families in the region. Yet, as with others in Schoemansdal, the ability to grow crops is shaped by family health and historical land-tenure systems that leave only a fraction of the population with space for agricultural production.

HIV LIFEWAYS AS SURVIVAL

The past decade has seen a remarkable transformation in the understandings and responses to HIV in South Africa. With widespread availability of ARVs, or what has been called "universal access" by SANAC, the possibilities for those infected have changed in profound ways. People are more willing to report their HIV status openly if they are on a treatment regimen that is helping to keep them alive. Some HIV-positive individuals note that they counsel others to get tested and treated if they have HIV. The infusion of capital from the national government and external sources has changed the landscape of HIV for those infected by the virus and created opportunities for dramatic reconfigurations of the domain of health in rural South Africa. Although the previous decade was marked by tremendous uncertainty and turmoil, the increasing availability of ART has meant that the country has entered an era of managed HIV.

Managed HIV is not chronic HIV because there remain barriers for some people to get tested and use ARVs. Additionally, adherence to ART is subject to behavioral changes and to accessing critical resources to ensure food security. As my discussion shows, managed HIV is the *possibility for survival*. Yet this possibility is neither certain nor guaranteed. It is the product of an individual's lifeway that positions them within a society while shaping their sense of self and confidence in the face of death. Managed HIV is survival, and this survival depends not just on access to ARVs but on a gamut of social and environmental resources that have become necessary to meet needs. The political environmental context makes people vulnerable to HIV, but it also generates the conditions for survival through expanded treatment options. The people and places described in this chapter are part of broader landscapes that produce differential vulnerabilities to HIV and possibilities for its management. The capacity to *live with HIV* is a product of spatial economies that have been produced over time by the convergence of political, economic, social, gendered, cultural, and ecological systems that create vulnerabilities and shape health and well-being. These landscapes are the material and symbolic realities for those now living with the disease. Individuals exist within broader landscapes of power that enable and constrain their health possibilities. Additionally, perceptions of health and opportunities for disease management are structured by these social and spatial systems. Health is spatial, and space produces health. This is the political environmental context that makes some sick and others well, while creating the conditions for some to get better while others do not survive. Our lifeways are produced by our spatial existence.

3 Historical Spaces and Contemporary Epidemics

HIV lifeways are produced by spatial processes that result in inequitable health possibilities. While the landscapes of HIV are shifting in response to changing governmental interventions, perceptions of HIV and wellness, and increased access to ARVs, it would be a mistake to read these as strictly contemporary phenomena. These landscapes and the places within them have been produced over time through the active imagining of space and its polity to benefit particular sets of interests. The spatial economies produced under colonialism and apartheid created an institutional structure that influences perceptions of health and the opportunities for decision making in response to illness. The historical construction of the Bantustans has material import for the ways in which disease and health are understood, experienced, and managed in the contemporary era. All three communities are contained within territory that was part of the KaNgwane Bantustan, and its construction was the result of centuries of spatial engineering.

The apartheid state built upon previous systems of racial classification and segregation established by colonial authorities, effectively leveraging legislation such as the Glen Grey Act that restricted land ownership by the indigenous population. British colonial administrators, as elsewhere on the continent, pursued the practice of indirect rule, whereby the creation of the native reserves under the management of tribal officers was the mechanism of control. In British colonial Natal, the British "diplomatic agent to the native tribes," Theophilus Shepstone, established the first African "locations" or reserves to serve as the spatial containers for the majority African population.[1] The passage of the Natives Land Act of 1913 extended the power of the reserve system by outlawing rent tenancy or sharecropping by Africans outside of reserve territories, while the Native Trust and Land Act of 1936 further consolidated these territories and the influence of tribal authorities within them. The latter act was central to pre-apartheid segregation through the expansion of the native reserve system and recognition of the tribal authorities as the legitimate governance system. The 1936 act worked to consolidate the reserves and would serve as the spatial foundation for the Bantustan system under apartheid. In order to acquire land, the local magistrate had to provide a permission to occupy (PTO) "to any person domiciled in the district, who has been duly authorised thereto by the tribal authority, to occupy in a residential area for domestic purposes or in an arable area for agricultural purposes, a homestead allotment or an arable allotment, as the case may be."[2] Secure land titling was not assured through this system, because PTO holders could be removed by the state even without payment, and financial institutions did not recognize ownership. This intensified the ability of the government to remove people by undermining the mechanisms of communal land title under private property regimes.[3]

With the election of the National Party over the United Party in 1948, the apartheid government extended the British system of indirect rule and aggressively intervened in constructing tribal governments and extending the power of the tribal authorities. The Bantu Authorities Act of 1951 recognized the tribal authorities as the chief governing system and abolished the Native Representative Council that was created by one of the 1936 "Natives" acts. Bantu authorities were organized into tribal, regional, and territorial levels, and at all levels the tribal authorities were dominant. The tribal authorities fell under the jurisdiction of the central government through the Department of Native Affairs, and the minister of native affairs had the power to remove any chief, cancel the appointment of any councilor, appoint any officer deemed necessary, control the treasury and spending, and authorize taxation.[4] The power of the tribal authorities was expanded during apartheid because the tribal authority had control over land alloca-tion. Coupled with spatial segregation policies that included the Bantustans, these were the only locations where the majority of Africans could legiti-mately claim land in rural areas. Members of the tribal authority could be rewarded by their complicity with the apartheid system, and Lungisile Ntsebeza has noted that ambiguities in traditional and contemporary tenure institutions enabled them to exploit villagers by charging for services such as state pensions, tribal courts, and migrant labor opportunities.[5] This would become one of the many contradictions of the apartheid system, because while tribal authorities were placed in charge of local government processes, such as the distribution of land or procedures for accessing certain resources, they were linked to the central government through the Department of Native Affairs, which gave the minister of native affairs direct control.

The Bantu Authorities Act was expanded when the Promotion of Bantu Self-Government Act of 1959 recognized eight "black national units," which formed the basis of the apartheid government's "separate development" strategy in the 1960s and '70s.[6] The construction of the Bantustans involved the classification of the national population into seemingly distinct ethnic categories and the forced relocation of people from both rural and urban areas. One of the few national surveys at the time, the Surplus Peoples Project, estimated that from 1960 to 1980 the proportion of the total black population living in the Bantustans, also known as "homelands," rose from 39 to 53 percent.[7] Figure 4 is a map of the Bantustans from the 1960s, after a period of consolidation when disparate territories were integrated. Often referred to as "grand apartheid," this was the period during which the initial apartheid ideology of the 1950s was institutionalized through formal policies that ensured spatial segregation of the majority of the country's population.

KaNgwane Bantustan was created in the 1970s through the establish-ment of the Swazi Territorial Authority, which had jurisdiction over the region. A legislative assembly was instituted in 1977, and in 1984 the national government granted "self-governing status" to KaNgwane. This was part of the apartheid ideology of "separate development," whereby the ten Bantustans were presented by national authorities as existing along a devel-opment trajectory, essentially a linear pathway that would conclude with their independence from the South African state. Separate development was

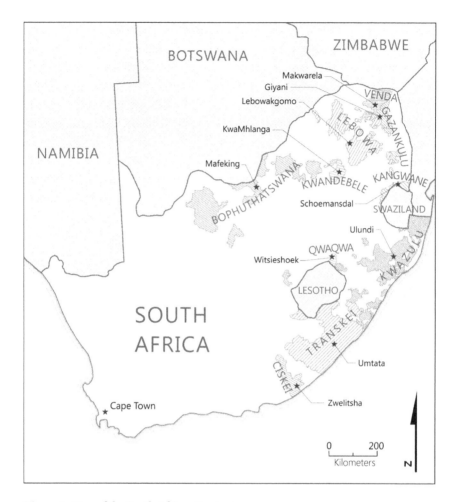

Figure 4. Map of the South African Bantustans.

the discursive strategy used to justify racial segregation, and it was advanced by invoking a paternalistic developmentalism.[8] State propaganda documents asserted that separate development was justified because the majority African population had the right to independent and self-directed development and that it was the responsibility of the state to facilitate this process. This strategy placed the national government in a position of trusteeship over the Bantustan territories, in essence justifying racial segregation while obligating officials to aggressively intervene in managing these landscapes.[9]

Propaganda documents asserted the benefits of spatial exclusion for rural Africans, with one declaring that "South Africa is recognizing the legitimate Black demand for self-determination, and is moving away from the essentially negative aspects of earlier policies towards a solution which accords each group its inalienable right to determine its own destiny and formulate its own scale of values."[10]

The construction of the Bantustans served as a critical element of apartheid's spatial economic system, ensuring a workforce near peri-urban and urban centers in the country. While my focus is on rural South Africa, the urban townships were also integral to these spatial economies in providing labor for manufacturing, mining, and service sectors in cities such as Johannesburg, Cape Town, and Durban. KaNgwane was the second-smallest Bantustan, established in 1976 with the creation of the Swazi Territorial Authority.[11] Regional authorities were established in Nsikasi, Nkomazi, Mlondozi, and Mswati, and townships were created within the territory for black resettlement from other locations. For example, people of Swazi descent living in White River were resettled in Kabokweni; those from Nelspruit in Lekazi; those from Matsulu in Kaapmuiden; and those from Malalane in KaMaqhekeza. The state reinforced the notion that local residents were advantaged by their proximity to fertile agricultural lands and were governed by traditional institutions to which residents were supposedly more loyal.[12] Government documents emphasized that KaNgwane was a separate country by listing its gross domestic product and national income statistics.[13] While advocating a separation between the Bantustans and the apartheid state, the landscapes of the Bantustans were in actuality the site of aggressive intervention by various state agencies intent on shaping social and environmental systems. The apartheid government reified the notion of eventual independence for these territories; however, the Bantustans maintained high rates of economic dependency and integration with the national economy. As evidence of this, in 1985 KaNgwane's gross national product was 1,260 rand per capita but its gross domestic product was 281 rand per capita—showing the dependence on work in adjacent areas.[14] KaNgwane was given self-governing status in August 1984, and four other Bantustans— Transkei, Bophuthatswana, Venda, and Ciskei—were classified by the national government as "independent," but this was resisted by some local leaders and was not recognized by the international community.[15]

With the dissolution of apartheid due to spatial policies leading up to the 1994 democratic elections, the boundaries of the townships and Bantustans were seemingly erased as these territories were reincorporated into the official geography of the country.[16] Regardless, this did not dissolve well-established sociocultural practices, institutional frameworks, governance regimes, and livelihood possibilities in the contemporary period. In fact, when the ANC government came into power, it was largely ambivalent toward traditional authorities in the rural areas, noting that they were well established and recognized by many of the country's citizens. Even though some traditional authorities had colluded with the apartheid state or had personally benefited by taking advantage of systems of land allocation and resource distribution, the newly elected leadership worked to draw a distinction between traditional authorities who were "genuine" and those who were considered "illegitimate." In the interests of developing political alliances with traditional authorities leading up the 1994 elections, the ANC began supporting "progressive chiefs" and institutions, such as the Intando Yesizwe Party of KwaNdebele, Inyandza of KaNgwane, the United People's Party of Lebowa, and the Bantustan regimes of Venda and Transkei.[17] KaNgwane's chief minister, Enos Mabuza, was the first Bantustan leader to visit the ANC and served as a model for other leaders who chose to ally themselves with the governing party.[18] Generally speaking, in the early years of the transition the ANC attempted to court the traditional authorities instead of articulating alternative governance models within the rural areas. A common feature of the relationship between the national government and traditional authorities is the assumption that the latter would concentrate on "traditional functions" rather than political or development activities. Of course, these distinctions proved difficult to maintain, and recent debates over strengthening the power of tribal courts underscore the complicated entanglements between traditional authorities and the ANC government in the post-apartheid era.[19]

SPATIAL LEGACIES AND LIVELIHOOD POSSIBILITIES

Rural livelihoods in South Africa are multifaceted and dynamic, reflecting histories of dispossession and territorial manipulation by the state. Colonial patterns of segregation, which were expanded through the Bantustan system in the apartheid era, produced social, economic, institutional, and spatial relationships that remain tangible, both materially and symbolically, in the contemporary period. One way in which spatial histories continue to be expressed is through the livelihood possibilities available to rural populations, which are integral to the maintenance of basic needs and for ensuring healthy lifeways. Livelihoods are generally understood as "the capabilities, assets (including both material and social resources) and activities required for a means of living,"[20] or the "everyday practicalities and diverse modes of making and defending a living."[21] Research on livelihoods has emerged from multiple fields, including agrarian studies, political ecology, and development studies,[22] with a focus on several themes, particularly capital assets, social relations and organizations, and access regimes that produce the opportunities available to individuals, households, and communities.[23]

The livelihood concept has an inherent materiality, so livelihoods are often examined as the possession and utilization of a diverse set of endowments or capital assets, which are categorized as natural, physical, human, financial, and social capital. As Frank Ellis has explained, natural capital consists of the stocks of the natural resource base (land, water, biological resources), while physical capital (also referred to as "human-made capital") includes the assets created by economic production activities such as infrastructure, tools, and agricultural technologies. Human capital refers to the education level and health status of individuals and populations, and financial capital includes stocks of cash or credit that are available for meeting regular needs or to provide investments. Finally, social capital, which has been increasingly recognized as integral to livelihood systems, is understood as the social networks and trust operating between individuals and communities.[24]

The ability of social actors to access capital assets for production is another critical component of livelihood systems.[25] Scholarship in political ecology and development studies has demonstrated the factors that result in differential access to resources, whether in terms of socioeconomic class, race, gender, or geographic location. In a notable contribution, Jesse Ribot and Nancy Peluso defined access as "the ability to benefit from things—including material objects, persons, institutions, and symbols,"[26] and while concerned with tangible resources, they emphasized the bundle of powers available to social actors to produce livelihoods by navigating political, economic, and cultural structures. In a similar fashion, entitlement frameworks acknowledge the role of endowments in alleviating poverty, while also emphasizing the structural constraints on how they can be leveraged. For example, a woman engaged in smallholder agricultural production faces different challenges in maneuvering external markets, and accessing credit from lending agencies, than a male farmer. Entitlement frameworks are derived from the work of Amartya Sen, who, in writing on famine in the 1970s, noted that poverty and hunger were a consequence not of the availability of resources, but of the application of these resources in specific contexts.[27] Central to this work, and to subsequent studies on environmental entitlements and capability frameworks, is the recognition that social actors are embedded in structural systems that simultaneously enable and constrain livelihood possibilities.

While livelihood studies have been foundational in many ways, there are limitations to how the concept has been theorized. First, livelihoods need to be understood not only as the possession and expression of material assets but also as the symbolic and cultural mores that are tied to institutional patterns and place histories. Since the ability to maneuver structures is differentially experienced by social actors, livelihood practices can vary by context. Recognizing this, Anthony Bebbington cautioned that assets were the basis of an "agent's power to act and to reproduce, challenge or change the rules that govern the control, use and transformation of resources."[28] It is also through the practice of livelihood making that individuals confront the limits of their agency and the boundaries established by external systems of power. These boundaries are established over time, and in South Africa as elsewhere, spatial processes generate differential political environmental contexts that shape livelihood possibilities and opportunities for health management. Whether in terms of securing economic resources for daily survival or accessing ARVs for managed HIV, livelihood systems confront the state and markets in profound ways. Second, some studies fail to attend to the reciprocal relationships between space and livelihood production. This means that space is understood as location rather than through the production of place that is created by the interplay between local actors, institutional interventions, state development discourses, and cultural practices.

Livelihood possibilities in rural South Africa have been created by centuries of racial classification and spatial engineering designed to benefit the minority population. The Bantustans were the location of aggressive intervention by the state to support agricultural, mining, and service sectors for the rest of the country, and while their establishment was justified through a paternalistic developmentalism, there were insufficient investments by the national government to support meaningful socioeconomic development. During the apartheid era, KaNgwane was presented by state documents as having a population reliant on agriculture and pastoralism, which resulted in the investment in these strategies while ignoring others that were routinely utilized, such as seasonal migration for employment.[29] Livestock were viewed as a key asset for livelihoods in the region and were valued "to only a limited extent commercially, but infinitely so on account of their religious and social significance."[30] National and provincial agencies justified their interventions in these areas as designed to counter cultural norms that, they claimed, resulted in overstocking of communal areas—thereby ensuring that livelihood practices were rational and scientifically based. A 1967 South African Department of Information report exemplified this approach in noting that "The Bantu peoples therefore own some of the best agricultural land in South Africa although they must learn to work it properly if it is to reach its maximum productivity."[31]

Regardless of the diversity of livelihood practices and high population densities in the former Bantustans, agriculture was promoted as a mix of dryland cultivation and cattle farming, and KaNgwane's land was deemed viable for "virtually any agricultural product."[32] The apartheid government consistently argued that agricultural land in the Bantustans was more productive than that in white-held areas, asserting that 100 acres in the former had, on average, the same potential as 147 acres in the latter.[33] In a similar manner to livestock management, development planners suggested that agricultural production would become "characteristic of developing countries moving towards the transition phase from subsistence farming to the production of a surplus" by supporting technical expertise to maximize production, improve crop information, and facilitate expanded access to markets.[34] These representations ignored the realities for rural populations, including the tangible constraints placed on agricultural production because of apartheid spatial planning.

As a result of these spatial histories, livelihood systems in the contemporary era are highly diversified, with some households fully engaged in formal economic employment while others depend on government support in the form of grants. In other cases, families rely on a mix of strategies, including small-scale farming next to the home, part-time employment, reliance on remittances from an external family member, and utilization of natural resources to offset costs and meet basic needs. Agricultural production outside of household gardens is uncommon because of the high population densities in the region, poor soil quality, and scarcity of water for irrigation.[35] Regardless of the sweeping invocations by state planners as to the viability of agriculture and pastoralism in KaNgwane, prime agricultural land was removed from African communities and granted to white commercial farmers. The consequence is that the former KaNgwane Bantustan remains surrounded by larger-scale agricultural production, particularly for fruit and sugarcane farming. Seasonal and permanent migration to urban areas, such as Mbombela and Malalane, remains a common strategy for community residents searching for employment in neighboring industries.

The reliance on traditional authorities as the mechanism of indirect rule has resulted in continued negotiations that produce livelihood possibilities in rural South Africa. In addition to land allocation for commercial or residential areas, the tribal authorities have historical claim over the communal areas that are adjacent to village boundaries. Communal areas are open spaces in which residents collect natural resources, graze livestock, and, over time, settle in demarcated housing plots. In speaking with provincial governmental officials, they are inclined to note that during and after apartheid these lands are under their jurisdiction. By contrast, the chief of the Matsamo Tribal Authority once confidently stated to me that this land was "his." Yet even with these inconsistencies, the communal areas are critical spaces supporting the livelihoods of rural residents. People collect a mix of natural resources from the communal areas; wood, sand, thatch grass, medicinal plants, and use of communal space for livestock grazing are the most important to household economy. Numerous studies have documented the variety of natural resources and their general contribution to household economy and livelihood systems.[36] Wood is routinely collected for cooking and building; sand is collected or purchased by a majority of households for building; and it is not uncommon for grasses, reeds, insects, and wild fruits and plants to be collected. Medicinal plants can also be collected or purchased for a variety of ailments, and while the use of these materials is often underreported, there is ample evidence in the scholarly literature of traditional medicine's importance to health management in South Africa.[37]

The absence of formal and informal economic opportunities in the region means that many individuals are actively seeking employment. Regardless, unemployment for youths aged fifteen to twenty-four years was estimated at 40 percent in 2012. The same report noted that "youth unemployment also reflects the spatial inequalities that stem from apartheid-era policies of 'separate development.' Using the broad definition, youth unemployment in rural areas is 82 percent compared to 58 percent in urban formal areas."[38] This contributes to migration to neighboring areas, including commercial farms in the region. Additionally, many families depend on government grants that are distributed by the South African Social Security Agency. These are provided for a number of reasons, including child-care support, old age, and disability. A recent report noted that roughly one-third of South Africans receive some type of social grant.[39] In numerous interviews conducted in the region, the importance of grants for household economy was emphasized as critical in meeting basic needs and providing for family members. Residents indicated that the grants were not sufficient and should increase; however, this system is keeping many families out of extreme poverty. As noted in the previous chapter, documentation is required for many of these sources of support, and given the proximity to Mozambique and Swaziland, where cross-border migration occurs, some residents have difficulty accessing governmental assistance.

While colonial and apartheid spatial planning have resulted in geographic legacies for rural communities, these landscapes are simultaneously being transformed in the contemporary era. A variety of development agencies are operating within rural areas, promoting irrigation schemes, sugarcane farming, and conservation and tourism initiatives that are being advanced through economic neoliberalism. Observing the shift immediately following the 1994 democratic elections, Chris Tapscott argued that "By shaping the pattern of regional development, and by determining the type of economic programmes which have been funded

in the bantustans, for instance, 'development discourse' has had (and continues to have) a material impact on the lives of millions of South
Africans."[40] The consequence is that the communal land surrounding the
three villages is being transformed in ways that directly affect households
reliant on natural resources for livelihood production, creating tensions in
the region over access to land for other development schemes. In one
example, a group of livestock owners from the Mzinti community convinced the tribal authority to set aside land for a cattle project. What was
notable was their invocation of pastoralism as a "traditional" cultural
practice to claim this territory, even though only a fraction of the community owns livestock.[41]

The importance of natural resources to individuals and families, taken
together with their location in spatial territories that experience conflicting and potentially contradictory governance regimes, demonstrates the
importance of the political environmental context in shaping the possibilities for livelihoods and health. While these landscapes are experiencing change in contemporary South Africa, social and ecological features
are persistent and remain meaningful for rural communities. As the next
section demonstrates, the collection and use of traditional medicine has
been a feature of livelihood systems in the region for some time. This has
been particularly important for health management; however, HIV/AIDS
has challenged perceptions of the efficacy of these practices and exposed
tensions between state preferences and local practices. In simultaneously
recognizing and regulating traditional medicine, the national government
has transformed the ways in which traditional medicine is understood and
utilized in rural areas. An examination of shifting discourses of traditional
medicine, particularly during the HIV/AIDS epidemic, demonstrates how
livelihoods and health management remain intimately connected to historical spaces and cultural practices.

TRADITIONAL MEDICINE AND MANAGED HIV

Narratives of health decision making that emphasize unitary categories of
medicine simplify the complexity of how bodily health and care seeking are
understood. In South Africa, reifications of "traditional" and "Western" med-

icine do not capture the reality that individuals negotiate between and within these different options for a variety of reasons. Although some existing scholarship emphasizes the regular use of African traditional medicine, gaps remain in our understanding of the means through which individuals pursue particular health-care options and also the sequencing through which clinics, medicinal plants, traditional healers, and hospitals are pursued.[42] Moreover, a plurality of healers are labeled "traditional," including *sangomas, inyangas,* and *umthandazis.*[43] Although other regions in South Africa draw distinctions between sangomas and inyangas, within my research setting they are perceived by residents to be essentially the same, with the exception that sangomas have received additional training to commune with ancestral spirits and provide counsel to patients. Umthandazis are faith healers who often use prayer and Biblical teachings as part of their caregiving. In addition to the multiplicity of options, perceptions of traditional medicine and other forms of care collide with social and cultural understandings and practices. Mandi, a woman in one of the focus groups, emphasized that African medicine was different from Western medicine because it could be drank and shared with no measurements regardless of whether food had been consumed. By comparison, she explained that Western medicine did not allow for this because it had to be measured. Continuing, she said that African medicine could not be used for HIV, but rather that medicinal plants and traditional medicine would be used when someone in the household was infected by an "African disease." Another participant chimed in to say that sometimes people die when using African medicine because there is no one to monitor traditional doctors. Should someone need blood or water in the body, it could be dangerous, but where one seeks out care depends on a preference for treatment interlinked with personal belief systems.

These expressions around traditional and Western medicine are consistent with earlier work that I have completed in South Africa. In one study, I evaluated the reasons that respondents gave for not seeing a traditional healer such as a sangoma. Fifty qualitative interviews and 478 structured household surveys conducted during 2001–02 were used to examine the reportage of traditional medicine use in the village of Mzinti and the motivations for pursuing particular treatment options, including traditional healers, clinics, hospitals, and private doctors. Six percent of household respondents in the structured surveys reported collecting medicinal plants, and 26 percent indicated that a family member had visited a sangoma for treatment. Interestingly, the 2013 survey showed consistent patterns in reportage of the use of traditional medicine. Across the three villages, 7 percent shared that a member of the household collected medicinal plants and 21 percent had a member who had visited a traditional healer at some point in time. Lastly, when asked about the use of traditional medicine in the community, more than half of the respondents indicated that some or most of the community used traditional medicine. This pattern echoes the reporting of HIV in which a reluctance to disclose the use of traditional medicine in a personal way is not matched by the higher estimates given for the community as a whole.

Supplementing these survey data with the semistructured interviews conducted in 2001 and 2002 provides a fuller portrait of the factors that inform perceptions of disease and decision making among care seekers. First, local residents consider certain illnesses to be "traditional" illnesses that then require treatment from traditional healers in order to improve health.[44] People speak of being bewitched by spirits or of needing assistance in interpreting messages from the ancestors, while others insist that traditional medicine can help if someone is arrested for committing a crime. One member of the men's focus group from Ntunda described the importance of traditional medicine in the community. He explained that

> in our homes we believe strongly in natural resources because as soon as a child is born, the child will be given *imbita*, African medicine, immediately when it comes out from the mother's stomach. Even if you get sick and the clinic will delay you from your flu, you will just take guava leaves and *linyatselo*, boil and drink it and become well.

In addition to traditional maladies, medicinal plants and traditional healers are pursued for treatment that could be addressed at the clinic or hospital. Mzinti residents shared that they visit traditional healers for various purposes, ranging from headaches to stomach problems, cancer, tuberculosis, and strokes. While adherence to traditional medicine is common in the study region, many people pursue traditional healers only after first visiting the clinic or hospital.[45] In attending to HIV symptoms, this is reinforced by residents. Traditional healers explained that they encourage prospective patients to first visit the clinic to be tested for the virus. Drawing upon information collected from the 2002 structured survey in Mzinti, I conducted an assessment of the reasons people elect not to visit a sangoma. In the bulk of cases, respondents indicated that religious faith was determinative in their decision making. For these individuals, being Christian meant adhering to the power of prayer and clinical medicine in the treatment of illness. This shows the power of belief, in these cases the faith in other powers to shape health and wellness. It is also worth emphasizing that people might choose not to share their visits to traditional healers because of their belief systems. Elinah, who was working as a research assistant with me at that time, explained that people were not comfortable sharing that they visited traditional healers. Laughing, she indicated that many in the community "pray at the church in the day and visit the sangomas at night."[46]

The use of traditional medicine can be seen in other ways throughout the landscapes of the study region, whether in the routine practices of resource collection or at the markets scattered throughout the villages. After a meeting with members of the Matsamo Tribal Authority, my research team visited a market that was being held in Schoemansdal. This is a weekly occurrence that stretches along the dirt road from the tribal authority's offices toward the main road that runs from Malalane and Swaziland. It was a bright, sunny day and quite hot even in late morning, but there were dozens of market stands being tended by residents from the area. The proximity to the tar road and the relative size of Schoemansdal compared to other villages in the area resulted in a vibrant setting, with many people walking from stand to stand or engaging in active conversations with the vendors. As we walked along the dirt road, the most prominent stands were those selling fruits and vegetables, cooking and cleaning equipment, or large bags of *mealies*, which is cornmeal that is mixed with boiling water to make the traditional porridge called *pap*. Interspersed with these stands, however, were others selling plants collected from the surrounding communal areas or produced by local sangomas.

As seen in figure 5, there were a variety of products being sold at this one stand. Bags of roots and barks circled the tarp spread out on the ground for display. Closer to the center were some prepared materials in plastic containers that sat next to items that would be used by sangomas in attending to their patients. Calabash gourds were placed next to shells

Figure 5. Medicine market in Schoemansdal (November 2012).

that would be used in "throwing the bones" to assist patients in communicating with their ancestors or interpreting the future. The conch shells are thrown, and, depending on how they land on the ground, they are held to the ear of the sangoma to indicate that good news might be coming to the individual in the future. Also on display were animal parts, such as impala horns and baboon skulls. Behind the tarp were assortments of medicines in plastic containers that would be dispensed to customers upon request. In past interviews with Mzinti residents, it was shared that when *muti* (traditional medicine) is purchased, it comes either directly from a healer at his or her home or from medicine markets such as this one. Roughly 8 percent of the households surveyed in 2013 reported purchasing traditional medicine in the market, which would be used for various purposes, including stomach pain, headache, bladder problems, and wounds. It should be noted that muti (also spelled *muthi* or *umuthi*) can also include

potions that can be used to inflict harm on another person, and is sometimes referred to as "witchcraft." As some of the writing on the subject has
detailed, much of the power of muti is ascribed to the belief in its effectiveness, meaning that the user must be convinced of its effectiveness or it will
not work.[47]

After inspecting the products on display, I asked the owners of the
stand if they had African potato. An elderly gentleman, a sangoma, hur-
ried to another stand and brought back two specimens for us to examine.
The African potato is a tuber with long roots resembling an octopus. One
of the specimens was recently collected from the surrounding communal
areas and was still light brown and moist. The other had been dried and
was darker in color. Many of the roots had fallen off the latter, and what
remained were wisps of frizzy hair. We talked with the sangoma about the
plant, which is prescribed as a tea that comes from boiling the tuber to
release its nutrients. Elinah shared that this could be dangerous and you
needed to know how to prepare the tea, because overboiling could make it
toxic. When asked if it was used in the treatment of HIV, the sangoma
explained that the African potato "prevents HIV."

I had been hearing whispers for years that the African potato was being
prescribed to HIV-positive patients, so it didn't surprise me to hear the
sangoma assert its medicinal properties. At the same time, during the
height of the national government's resistance to establishing a national
ART program, much consternation was directed by the global health community toward the assertion of the treatment value of traditional medicines. This locates these medicines within conflicting discourses about
what is beneficial for managing human health. These understandings collide with others, revealing competing views held by different political and
cultural institutions. The statement that the African potato prevents HIV
has multiple meanings. From a Western biomedical perspective, it would
be interpreted as false because the African potato cannot stop the spread
of a sexually transmitted disease such as HIV from one person to another.
But if some of the research on syndemic interactions is accurate, specifically that a healthy immune system plays a role in reducing vulnerability
to transmission, then this statement has a different meaning.[48] African
potato extract has been shown in clinical studies to have nutritional

benefits and high antioxidant efficacy compared with other extracts such as olive leaf or green tea.[49] So while the temptation might be to interrogate the sangoma's statement as either true or false, the reality lies somewhere between the two. For many residents, perceptions of the efficacy of the African potato operate within their lived worlds and are shaped by their beliefs and immediate needs. Classifying perceptions of health decision making as either true or false obscures deeper references that people ascribe to their well-being. The multiple meanings attached to HIV/AIDS, and the different ways it is understood and managed, show that in South Africa it is a distinct disease.

Conflicting interpretations of HIV/AIDS within South Africa and their role in producing health discourses was also evident in the responses by governmental officials challenging neocolonial or development language. During the early years of the epidemic, ANC leaders emphasized this by blaming Western governments and pharmaceutical companies for benefiting from the crisis. Additionally, they were concerned with the behavioral perspective that attributed the spread of the virus to irresponsible decision making as opposed to structural factors such as socioeconomic poverty. As South African scholar Peris Jones has detailed, the HIV/AIDS epidemic in South Africa at that time was

> bound-up with far more than merely a battle of organisms and biology. Rather, the epidemic has also been associated with prior cultural understandings of what induces vulnerability to the disease, including a prescribed pathology of certain groups. HIV/AIDS therefore exacerbates existing stigmatisation and exclusion directed at less powerful groups and individuals.[50]

Disease reveals asymmetries of power and discursive battles that are embedded in the underlying structural determinants of human health. In contrasting the differing HIV prevalence rates and state responses in Uganda and South Africa, Robert Thornton asserted that Uganda had benefited from attempts to "indigenize" AIDS so that the epidemic was understood in terms other than the biomedical. He suggested that the coordinated state–NGO response was part of a nation-building endeavor that linked the disease to indigenous understandings of cultural forms and knowledge.[51] By contrast, Thornton argued that South Africa's aggressive attempts to obfuscate the disease while delaying the release of ARVs had also shaped how the epidemic was understood, but in a markedly different way.

These histories produce the political environmental context that individuals navigate in order to understand their health and well-being and to seek out livelihood possibilities to manage illness in the contemporary period. This context changes over time but is interpreted through localized social and cultural practices that filter the interpretations of healthcare options. The perceptions and use of traditional medicine intermingle with understandings of decision making around health in complex ways that are shaped by historical, cultural, gendered, and religious conventions. Additionally, in the era of managed HIV, there has been a hardening of perspectives about the efficacy of traditional medicine. Consistently it was stated that when someone is "sick," meaning symptomatic of HIV, they should get tested at the clinic or hospital and then receive ARVs. Across all six of the focus groups, people drew a distinction between the use of traditional medicine and conventional drugs for the treatment of HIV. Generally, HIV was seen as untreatable by traditional medicine and needed to be dealt with at the clinic or hospital. For example, a woman from Schoemansdal stated:

> If you have HIV the sangoma is wasting your time. You will die because the inyanga will give you something to make you vomit and say you have this . . . but it is not like they know you have the HIV virus. It does not allow you to vomit if you have the virus. The inyanga cannot tell you that you have the virus . . . so in the hospital they can see you very fast because they will inject you and take your blood and then if it is positive they will give you pills immediately that you can take and eat and become healthy again.

The political environmental context is subject to political and bureaucratic interventions that are justified through discursive fields of regulation. Perceptions surrounding traditional medicine are not strictly individualized or local, but are also national and global, and are filtered through cultural and political economic systems. This has certainly been true for African traditional medicine, because the national government has been active in simultaneously reifying its use while attaching regulatory expectations to healers and their knowledge base.

RECOGNITION AND REGULATION
OF TRADITIONAL MEDICINE

Given the fluidity with which people pursue care from clinics, hospitals, private doctors, and traditional healers, the national government has worked to institute policies that regulate traditional medicine. In 2004, the government passed the South African Traditional Health Practitioners Act 36, which prohibited traditional healers from diagnosing or providing treatment to patients with HIV/AIDS, cancer, or other terminal illnesses. The Constitutional Court later found the act invalid; regardless, the health minister at the time, Dr. Manto Tshabalala-Msimang, worked actively to integrate traditional medicine into the national health-care system. The Traditional Health Practitioners Act (No. 22 of 2007) established the Traditional Health Practitioners Council with the intention of overseeing traditional healers. The Policy on African Traditional Medicine in South Africa, drafted in 2008, stated the need for official recognition, empowerment, and institutionalization of African traditional medicine.[52] Stating that many South Africans rely on traditional medicine for routine care, this draft policy advocated its use while insisting on its regulation by the government. The document is also notable for emphasizing that national mainstreaming and regulation need to be centered on "an evidence-based public health and epidemiological approach, supported by laboratory-based investigations."[53]

Interviews conducted in the communities evidence a seemingly rigid discourse surrounding traditional medicine for health complications, in particular regarding HIV/AIDS. Time and again, respondents emphasized that when an individual becomes symptomatic of HIV, they need to get tested to determine whether they have the virus. They noted that in cases when patients first visit a traditional healer, they should be instructed to get tested before undergoing any form of treatment. One man explained that the government doctors and traditional healers were now working together in addressing health challenges, especially HIV and tuberculosis (TB), noting that traditional healers will call other doctors to obtain information about how to treat them. Continuing, he said that

> I think that inyangas are no longer practicing like in the past. Now they are educated in our health because really there are diseases that need natural resources to cure them. So these diseases you go to the inyanga and get med-

icine and get healed because there are also diseases that you cannot be cured from unless applying traditional medicine.

A woman from Schoemansdal stated that "in some ways nature can help but in the cases of diseases like HIV or TB you cannot use nature. These diseases do not want nature. Like it or not, you have to go to the hospital." In still another example, a lively exchange occurred during the Ntunda women's focus group after one of the participants shared that she was a traditional healer. Elinah, who was facilitating the discussion, intervened to ask several pointed questions, but the healer was insistent that HIV is not curable and should not be treated by traditional medicine. She explained that when a patient visits her saying that they are ill, she insists they go to the clinic to be checked to see whether it is only TB or whether the patient is also HIV-positive. She noted that the patient

> must come with the results and show me to prove that they are sick. Because I am also a doctor, so that we could be able to continue healing each other. You would tell me lies saying you are right while I am busy treating you, saying you are suffering from running stomach and to find that you are HIV-positive until such time that you die. Sometimes you are talking lies [about visiting the clinic], claiming that you did go there, wasting your time here and end up dying in front of me.

This healer was adamant that the patient needed to get tested to determine their HIV status before they could make an informed decision about their treatment options.

While other members of the survey team were out administering the questionnaire, I had a conversation with Nelson, who was waiting to begin his work for the day. In discussing his views on the generational differences between the three villages, he stated that the elders believe that eating certain things from the bush, such as a wild fruit called *intoma*, is very good for your health. He commented that the elders would say it is a hundred times better than an apple. Nelson pointed to a plant growing at the business center and said that they would take the roots from that plant and eat them. When asked whether younger residents believe this he said no, that they would prefer to visit doctors. At this point he got more animated and began talking about the importance of believing in the effectiveness of treatment practices. "If you take an aspirin for a headache and it goes away in twenty minutes then it is your brain telling you that you are feeling better. Not the medicine." Nelson then mentioned the moringa tree, which people in Nkomazi were talking about as being good for health. I shared that I had heard of this and that it boosts the immune system, remembering a conversation with two of the directors of a local resource center who had noted that there were development projects in the area to encourage people to plant the moringa tree. One of these is being promoted by Trans African Concessions, which built the N4 toll highway that connects Johannesburg and Pretoria to Maputo in Mozambique. This project is working to provide fifteen thousand trees in the area at a cost of one dollar per tree. One of the directors told a story of having a pickup truck filled with harvested leaves and branches from the tree and finding that no mosquitoes would come near, indicating its usefulness for malaria prevention. She also noted that moringa can provide an immune booster for people who are HIV-positive. Nelson agreed with this assessment and said that some people were even saying that it cures HIV. He explained that the leaves are dried and can be eaten or boiled and consumed as a tea. The consequence for some was that "they are sick and eat from the tree and they are better." Always thinking like an entrepreneur, Nelson indicated there would be a "demand" for the moringa tree in two or three years because people were becoming increasingly aware of it. He said people from Swaziland and Mozambique were coming, and might come in the future, to Nkomazi for the moringa tree. Nelson smiled in saying that if you asked the elders they would say they always knew about the moringa tree, that this was knowledge they had all along.

It was uncommon for respondents to discuss the use of traditional medicine in the treatment of HIV, though they might report using it for associated symptoms. This was also indicated in some of the qualitative interviews I conducted in 2014 and 2016. Respondents were adamant that they did not "mix" ARVs with traditional medicine, explaining that this would reduce the effectiveness of these drugs. Residents also shared that they were advised by the clinics where they received their ARVs not to visit traditional healers for any reason. This demonstrates a regulatory complex whereby the state has emphasized a particular discourse and set of practices for HIV management that is associated with ART. While national documents assert the benefits of aligning traditional medicine with other forms of health care, clinics and hospitals are insisting on a specific regimen for managed HIV. This is expected to produce another form of behavioral change to accompany adherence to ART. Beyond HIV/ AIDS, what also appears to have shifted in recent years is the protocol by which people first seek out care, and the ways in which traditional healers are integrated into Western clinical facilities. Respondents insisted that people would go to the clinic or hospital first and then go to the traditional healer afterward. Some informants indicated that they thought the tradi- tional healers have gotten strategic in first referring patients to the clinic or hospital. As one participant in the Schoemansdal women's focus group noted:

> The inyangas have seen that before they can assist you they will take you to the hospital. The reason is that they are running away from those that have TB or HIV. People have died because of that. Now they are clever and real- ized that the first thing is to go to the hospital. Then after, you come to them. The inyanga will go with you to the hospital and if you have the virus they then leave you there for treatment until you are well. They no longer do what they used to do in the past because people were dying.

While I can only speculate on the degree to which these views have changed since the first documented cases of HIV in the country, I can reflect on some of the observations from my fieldwork in Mzinti in 2001, a time when people were reluctant, if not completely unwilling, to disclose their HIV status. For example, I conducted a lengthy interview with a san- goma. Samuel was a young man, in his thirties, who was quiet but had an energetic demeanor. The interview was conducted in a second building, detached from the main house, where he practiced medicine. Along the walls he had bones from animals such as pig, impala, and duiker (a very small species of antelope). These bones were purchased at the local market or Samuel had collected them from the bush when he was out gathering medicinal plants. Many medicines were contained in liquor bottles, with dozens stacked on a shelf on one side of the room. While some were liquids, he also had a bottle containing powder from snakes, and a snakeskin believed to help with pregnancy. He proudly showed me a hat that he had made from a duiker hide, with African wildcat fur on the top. Samuel was born in Mzinti before moving to Steenbok, which is a smaller community near the Mozambican border. He had recently returned to Mzinti to be closer to the services in the area, in particular to have a steady supply of water, which, he emphasized, was necessary in the treatment of his patients. The discussion of water brought up the subject of HIV and he shared that he was trying to collect certain herbs to treat boils associated with the disease. When asked if he was treating anyone for HIV/AIDS he said that he was not, though he was seeing people who were clearly symptomatic. In cases in which he considered himself unable to treat the patient, for example if they had a shortage of blood, he would refer them to the clinic or the hospital. As he noted, "I am not a selfish person or either curious to do something like I know everything. I know that if I take a chance and that person dies, I will sustain my destruction from God."

To draw a contrast with this interview from 2002, it is instructive to include another exchange that occurred with an umthandazi from Schoemansdal in 2013. Members of my team spoke with Peter because he had disclosed the previous week, during the household survey, that he was HIV-positive. In telling my team about his past, Peter shared that he had learned about traditional medicine while serving as a teenage apprentice to his mother, who was training at that time to be a sangoma. During those training sessions he learned about the different trees and plants that could be used for treating patients. One night he had a dream in which two trees appeared to him from which he learned how to make muti. While not formally trained as a sangoma, Peter explained that he prayed with his patient and placed his hands upon those who were suffering. During this process, spirits would come to him and tell him about the person's affliction so that he could provide proper treatment. He emphasized the importance of this because in some cases people were bewitching others or even, during springtime, sending lightning and thunder to harm others. Because of his former employment as a driver in Mbombela, Peter collected muti from that area in 2009 to use in treating patients. Although many of the medicines were in labeled containers, there were others that he didn't want to be recognizable, so he kept them in Coke or Fanta bottles for concealment. Peter explained that some of these medicines are dangerous and so he kept them hidden from his children for fear that they might try to use them when he was not home. He was currently taking ARVs from the clinic, though he noted it was difficult because the number of pills was always changing and they were expensive. When asked how the medicine made him feel, Peter insisted that he felt good and strong. He then removed his shirt to demonstrate this to my research team. As with other people in the area, the virus was present but experienced differently than in the past. People taking ARVs embody the disease in new ways that were previously not possible. The domain of health for Peter has shifted because of changes in the political environmental context. It is expressed in novel ways, suggesting an emerging lifeway with a more hopeful future.

4 Landscapes of HIV

An article published in 2012 in a local paper, the *Corridor Gazette,* is instructive in outlining some of the dominant discursive framings of HIV/AIDS in contemporary South Africa. Titled "Nkomazi excels in the fight against HIV/AIDS," it reported on a meeting held in Kamhlushwa with local AIDS councilors and "Member of the Executive Council" (MEC) for Health and Social Development Dr. Thamiord Mkasi.[1] During the meeting, Dr. Mkasi commented that the Nkomazi Local Municipality was a leader in the fight against HIV/AIDS.[2] The paper indicated that what set the campaign apart were the foot soldiers who comprised a group called Total Control of the Epidemic that were responsible for "fighting the battle." In describing what made their response distinctive, the project supervisor explained that the group works actively by going door-to-door to provide testing and counseling to potential patients in their homes. As he noted, "We only visit people when they call us, we don't just go into their homes and start performing HIV/AIDS tests, they call and make appointments." This approach highlights the newly aggressive stance of state responses to the epidemic, whereby residents are approached by those on the front lines who surveil the population to provide testing and dispense educational materials.

A second news report published later in the year highlighted yet another element of the public health campaigns. In the article, the Department of Health noted the importance of a massive campaign to create much-needed awareness of HIV. To support its efforts, the department had recruited the newly crowned Miss Mpumalanga to conduct awareness campaigns throughout the province. Another MEC, Ms. Candith Mashego-Dlamini, stated that while the "HIV scourge" was a national problem, Mpumalanga Province's HIV prevalence was the second-highest after KwaZulu-Natal. Through the Zazise ("know your status") campaign, the Department and Miss Mpumalanga were focused on reducing the pandemic by at least 50 percent by 2016. Speaking of her role, Miss Mpumalanga noted that "I want to focus mostly on women and young girls, who are often taken advantage of because of their backgrounds. I hope that young people would be encouraged by other young and successful people like myself and we hope they would want to start leading a positive life."[3] In the same article, Miss Mpumalanga also stated that "HIV/AIDS may be a killer, but we will show, together with the MEC, that we are conquerors as we will be saving a lot of lives."

Military metaphors circulate in other venues. Consider a discussion document that accompanied the notes from the April 2005 meeting minutes of the Mbombela Council Chamber of the Ehlanzeni District Municipality.[4] The central item on the agenda was an evaluation of provincial efforts to address the epidemic while engaging in a discussion of future coordination efforts. Particularly noteworthy was a discussion document accompanying the minutes that included a section from the HIV/AIDS/STD Strategic Plan from 2000–05. Likening AIDS to armed combat, it stated that the pandemic has led to more deaths than all wars combined. It continued by noting that

> these disorderly conflagrations were regarded, quite correctly, as extremely destructive because they came with a lot of noise, fire and smoke than is the case with the AIDS pandemic. . . . AIDS on the other hand, is neither noisy, nor is it full of smoke and fire, not does it overtly and directly destroy the means by which humans sustain their lives; it is rather a very silent and insidious killer that is more destructive, in human terms, if not the most destructive force in the entire human history.

The rationality and agency through which the epidemic is described in the Strategic Plan is striking. HIV and AIDS are emphasized as "the most pervasive and cunning enemies that mankind has ever faced." The virus is described as a "wily strategist" and AIDS as "its destructive relative" that locate themselves in a strategic attack position within a human's bodily fluids. The document identifies AIDS as a threat to the very survival of the human species that can only be attacked through a social movement that mobilizes "all people to act in unison in their titanic fight against this common deadly enemy."

The invocation of military metaphors is revealing and highlights the perception that the population is under siege and that equally aggressive responses are needed. A war is raging, and soldiers are needed to win the battle. Yet not unlike military battles, public health campaigns are fought upon landscapes that have their own histories and memories. These landscapes are the spatial context through which infectious diseases spread, and they generate the opportunities for health management following transmission within the population. Public health campaigns are also discursive battlefields through which particular understandings of human health become hegemonic at the expense of competing understandings. Campaign messages often collide with localized, gendered, and cultural norms and practices that filter out the messages in revealing ways. Attending to the dominant health narratives that are portrayed by governmental representatives, media outlets, billboards, and other public displays reveals multiple layers of the health landscape, not all of which speak in unison.

My intention in this chapter is to uncover the overlapping layers of HIV landscapes to show how they have been produced over time through the interactions between social and ecological systems. I do this in order to demonstrate the ways that South Africa's HIV/AIDS epidemic is distinctly South African. While it bears the signature of the disease in other contexts, it has particular features that have been produced by society and space during the colonial and apartheid periods. The political environmental context contributes to explaining the disparate vulnerabilities to HIV infection, competing understandings of the disease and potential ways to manage it, and the multiple impacts it has on those infected and affected. This has produced landscapes of HIV that social actors must navigate in order to avoid transmission or to access ARVs and other resources needed for survival. While these landscapes are spatially expansive and constraining, they are also fluid, depending on the ability of people to maneuver them.

LANDSCAPES OF CARE: ENCOUNTERS
BETWEEN THE STATE AND CITIZENRY

The complex terrain navigated by those infected and affected by HIV is produced by political, economic, gendered, cultural, and ecological relationships that shift over time and space. These landscapes are produced through the political environmental context that generates differential vulnerabilities to disease and poor health, uneven access to services and ARVs, and conflicting health discourses. The concept of a landscape has received considerable attention within the social sciences and has been a theme in geographic scholarship since the discipline's origins. Geographer Carl Sauer pioneered the idea of a "cultural landscape" to refer to the ways in which cultural practices shape the physical environment in observable ways.[5] Recent studies have concentrated on the co-constitution of labor and landscape, and on landscapes as lived and embodied by human populations.[6] These studies demonstrate the benefits of understanding space not simply as a location but in terms of the complex dynamics that produce distinctive places and landscapes. As noted by Melissa Leach and colleagues, landscapes can "come to embody layer upon layer of the legacies of former institutional arrangements, and of the changing environmental entitlements of socially differentiated actors."[7]

The landscapes of care generate distinct encounters between individuals and state institutions that shape the possibilities for healthy decision making and well-being. Access and use of state health-care facilities are not simply exchanges between particular individuals but are mediated interactions between the state and its citizens. For those managing HIV, or for other ailments and needs, these are the sites through which the state and its bureaucratic infrastructure are experienced. These encounters can be understood through theoretical analyses of how state power determines the lived experiences of its subjects. Michel Foucault's concept of "governmentality," which has been described as the conduct of governing, refers to the ways in which institutions of power and the social actors within them behave. Integral to governmentality is the power vested in the sovereign, the singular entity with the ability to decide on the exception—in essence, who lives and who dies.[8] This builds upon Foucault's notion of "biopower," which constitutes a shift from the historical mechanisms used by the state to regulate its citizenry. Rather than waging war and collecting taxes, the modern state governs through regulation by means of technical bureaucratization. Improvements in technologies of data collection and surveillance facilitate the modern state's capacity to know its subjects in intimate detail. As James Scott described, this is undertaken by forms of legibility, as in freehold tenure and standardization of language; and by forms of organization such as geometric design and urban planning that make it easier for the state to simplify and universalize its vision of the populace.[9] Through these technologies the invisible masses that were previously governed through the state's ability to execute legal violence became legible in ways that reduce the need for the state to exercise authority in this way. Biopower is routinely summarized as a shift in the state's role, from letting people live and making them die to making people live and letting them die.[10] Central also to Foucault's concept of governmentality are the ways in which individuals and populations internalize the logics of the state and regulate themselves to these expectations. This self-disciplining is the height of state power, wherein its policies, rules, and rationales do not need to be enforced through the use of violence because they are accepted by the citizenry.

These theoretical interventions have been productively leveraged in addressing HIV and health management. In describing the ways in which access to ARVs was negotiated in West Africa, Vinh-Kim Nguyen introduced the concept of "therapeutic sovereignty" to describe a novel form of political power in which the informal and formal procedures that determine who lives should not be interpreted as "technical, medical, or humanitarian issues. Rather, they are mechanisms that decide exception in matters of life and death."[11] At the time of his analysis from 1994 to 2000, Burkina Faso and Ivory Coast had scarce resources to respond to the growing epidemic, so acquisition of these drugs required navigating particular institutions and social relationships. Nguyen suggested that this was largely achieved through the act of testimonial, in which certain

individuals fitting an imaginary conceived by state and nonstate organizations were able to position themselves in proximity to those who could
dispense life-saving drugs. Nguyen argued that this created a "therapeutic
citizenship," in which the intimately personal nature of treatment programs colluded with responsibilities to practice sex in safer ways so as not
to infect others. Given the severity of the epidemic and the desperate need
to secure ARVs, therapeutic citizenship instigated new forms of political
practices and embodied subjectivity needed for survival.

In another assessment, medical anthropologist João Biehl differs from
Nguyen in seeing the negotiations in acquiring ARVs as necessarily a form
of top-down control. Drawing from ethnographic fieldwork in Brazil,
Biehl suggested that the state is not using therapies and grassroots services as governance techniques but as a political game of self-identification,
arguing that "proxy-communities, often temporary and fragile, and interpersonal dynamics and desires are fundamental to life chances, unfolding
in tandem with a state that is pharmaceutically present (via markets) but
by and large institutionally absent."[12] The varied and complex social institutions through which HIV-positive individuals maneuver underscores
how health is coproduced through these encounters. To Biehl, people are
agents in negotiating health outcomes, and even when highly constrained
by the restrictions created by social structures, "these configurations are
constantly constructed, un-done and re-done by the desires and becom-
ings of actual people—caught up in the messiness, the desperation and
aspiration, of life in idiosyncratic milieus."[13] The important point here is
that the relationships between the state and its citizenry are not necessarily clean or hierarchical. They are not inherently a global politics of
antiretroviralism but a more local form of negotiation between those who
have ARVs and those who desperately need them that result in a "highly
privatized politics of survival."[14] These relationships are formed by con-
tacts between social actors spread over dispersed networks that can
involve social movements, challenging the hegemony of state discourse or
reinvention through new public health campaigns.[15]

In a similar fashion, South Africans managing their HIV engage with
the state through health-care facilities that are unequal and spatially
dispersed, providing disproportionate advantages for some and inequitable options for others. As Didier Fassin has written, these are not just

encounters about disease and health.[16] Rather, he asserted, what is at stake is *survival*. Survival is the ability to secure scarce resources to generate monies for transport to a medical facility. Survival is about disclosing or hiding one's HIV status, out of preference, necessity, or fear. Drawing from Derrida, Fassin explained that survival is an existence between life and death, a state where life has "death for [its] horizon but which is not separated from life as a social form, inscribed in a history, a culture, an experience."[17] This is a meaningful intervention for at least two reasons. First, it suggests that individuals do not exist in a state that is either healthy or diseased; rather, their lived experiences and political circumstances result in an embodied position between disease and health. It is a fluid and dynamic state that is constantly unfolding. Second, Fassin's identification of survival is imbued with multiple meanings that are political and cultural but also spatial. Our existence as social beings is shaped in part by the processes that construct the places and landscapes of our world. How people navigate these landscapes in seeking out health care and other services is a product of historically rooted spatial processes that generate the opportunities and constraints to health and well-being. Factors that influence disease transmission in the contemporary period, whether they are infrastructure, gendered and cultural norms, or absence of employment in rural areas, have been produced over time and are the outcome of social relationships and power dynamics that have been spatially produced. The political environmental context is shaped by these historical spatial systems and mediates vulnerabilities to disease and possibilities for managing HIV.

Three key sites of health care exist in rural South Africa: local clinics, home-based care groups, and regional hospitals. Taken together with traditional medicine and healers in the area, which were discussed in detail in the previous chapter, these are the sites individuals navigate in responding to illness and managing their health. HIV lifeways are shaped by access to capital resources to pay for transport, should the distance be too great or not viable for walking; generational and cultural understandings about illness and the viability of either clinical or traditional care; and general preferences about particular options. This can be seen within the study region through the types of institutions that are available and how residents utilize them.

During a visit in November 2012, the Ntunda *induna* (the tribal authority's local representative) spoke very proudly of the clinic that was being built in the community, intended to open in 2013. When I returned in June 2014 it was still not completed; some residents explained that the reasons were unknown. It finally opened, with much attention, in 2015. Residents of Schoemansdal can also access a larger clinic in Buffelspruit, the village directly to the north; or a clinic in Jeppe's Reef, directly to the south. There is also a clinic in the Shongwe Hospital, which is located in Schoemansdal. Farther up Route 570 is the clinic in Malalane, which some consider to be of better quality than certain local ones. Should an individual have access to economic capital, they might take public transportation to the MedClinic in Mbombela for private care. It should be highlighted that the various clinics have different hours of operation. The Buffelspruit clinic is notable for being open twenty-four hours, and the Ntunda clinic is moving toward this, while the other facilities have more restricted operating times.

Mzinti, Ntunda, and Schoemansdal all have home-based care groups that are responsible for providing daily caregiving to community residents. These groups are staffed primarily by female community members, some of whom are paid while others work as volunteers. In speaking with representatives of these groups, they shared that their objectives are multifaceted. The type of care provided varies according to need and can include routine visits, assessing the condition of the household, and supporting the provision of drugs for treatment. They complete house visits to check on patients, either bringing patients to the clinic or hospital for drugs and other treatment or providing end-of-life hospice care for those beyond other options. Members of the home-based care group encourage people with children to attend the clinic, and they provide referrals to social workers and relevant departments such as the South African Social Security Agency. Additionally, they engage in public health and well-being activities such as promoting gardening projects to help people access food. Home-based care representatives talk about the need to test for all types of diseases and discuss hygiene practices. In going door-to-door, they are able to identify other patients who need certain services while also working with the clinics to track defaulters. People who are HIV-positive are told to use condoms to reduce the chance of transmission to others.

The work done by these groups is remarkable, especially considering the extremely limited resources available to them. Such reliance on local groups is a hallmark of the South African governmental response to HIV. This decentralization of care down to local clinics, hospitals, and community groups is in keeping with the government's embrace of neoliberal policies that push downward the responsibility of care and the sites of caregiving. The consequence is that for many, their primary engagement with the state is through their interactions with ward councilors, tribal authority officials, and health-care workers. The relationship between home-based care groups and residents is fluid and often dependent on external support. In recently speaking with one of the directors from a facility in Schoemansdal, she noted they had shifted to small and medium enterprise activities because the government had concentrated on funding only a few providers. A subsequent visit in 2014 revealed that the organization had fully committed to a tree project, working in conjunction with Trans African Concessions. When there was more of a financial incentive in the past, or a "carrot" as it was put to me, four home-based care groups had sprung up in Schoemansdal alone. Now it appears that home-based care is moving to the villages, and some of the groups are funded not by the South African government but by the European Union or the U.S. Agency for International Development (USAID).

The two hospitals in the region are the Tonga and Shongwe hospitals. Like many facilities in rural South Africa, they have comparatively impressive infrastructure and equipment though they struggle to retain staff. For example, while the Shongwe Hospital is a large facility, it has wings that were built but not opened because of the lack of personnel to support them. This is consistent with reports from other regions in the country. Citing Coovadia et al. (2009), Scrubb argued that overcoming inequities in human-resource distributions between the public and private health-care systems remains "a major existing problem in South Africa."[18] This is expressed in local capacity, because "health care professionals in rural clinics tend to have more poorly developed skills and less management experience, and the entire South African public health system is plagued by understaffing and overcrowding of patients."[19] The women's focus group from Schoemansdal was openly critical of the Shongwe Hospital, emphasizing that it was not well managed and that patients were not given the services they needed. As one participant explained:

You will find that the patient sleeps in plastic. The sheet has just rolled away, left in plastic. So you have to come with a sheet, blanket, and a fan because they are not there. Just imagine carrying a fan from over there to the hospital so as to provide the patient.

Another participant said it was necessary to bring water and soft porridge to the hospital and make sure your relative is clean even though they are in the facility. Regardless of these criticisms, interviews with residents indicate that the two hospitals and local clinics remain the primary venue for medical care and therefore are essential sites in the landscape of managed HIV.

HIV LANDSCAPES

Understanding the presence of HIV requires reading the landscape for signs and markings that indicate its presence. The messages, images, and meanings contained in the information communicated by state and non-state institutions convey much about the ways in which disease and health are understood. One of the few such billboards that could be found in the study region was in the village of Tonga, next to the main intersection with Mzinti (figure 6). The local municipality office is across the street, as is the police department. There is a public taxi stand behind the billboard and a newly built shopping mall, so it is a bustling center of activity.

On one side of the billboard is a presumably heterosexual African couple, dressed in white shirts and smiling at the viewer, who is implored to get tested for HIV along with any sexual partners. There is a communication of urgency in stating it is time to get tested, yet this is tempered by their relaxed and affectionate demeanor. The image suggests they have power in shaping their health together. Also noticeable is an SMS text number that when contacted did not respond, indicating that perhaps the public health campaign had ended. On the other side of the billboard is a second message, advocating male circumcision. A smiling male patient is being consulted by a smiling African doctor in clinical hospital clothing, seemingly preparing him for the procedure. This public health display emphasizes that circumcision can be done for free and advertises a number to text for more information. A third billboard, located in Schoemansdal, also promotes male

Figure 6. Billboard in Tonga: "The Time Is Now" (November 2012).

circumcision by appealing to its audience to "get the love cut this summer" (figure 7). This billboard is being promoted by the South African National AIDS Council, the Department of Health, and Brothers for Life. Located on the corner across from a major shopping center, there is consistent foot traffic and roadside vendors, ensuring prominent attention to its message. These signs are strategically placed at major intersections to ensure their visibility.

Male circumcision is a global message made local that attests to the individualized narrative of survival. Recent clinical trials have concluded that the risk of male-to-female transmission is reduced when the man has been circumcised, which resulted in the WHO advocating male circumcision in the management of the epidemic in Africa.[20] This generated a push for circumcision in the Global South as a means to reduce transmission, but it is not without consequences. The responsibility for managing the disease is located at the site of the individual male body, which is then

Figure 7. Billboard in Schoemansdal: "Get the Love Cut This Summer" (June 2013).

encouraged to submit to a Western clinical procedure, as opposed to the still practiced traditional type, in order to ensure physical safety. One member of the Ntunda focus group who had worked with the home-based care group was highly vocal on this point. During the discussion he pleaded with members of the group to get medical circumcision and was willing to organize the logistics for those interested, stating:

> I ask all of you guys to go for male medical circumcision. We must all circumcise to minimize the effect of this virus called HIV. As I'm talking to you I have been circumcised. You can come to me for free and I will register you and take you to one of the sisters at Kamhlushwa. I can call transport to come to you, as long as you reach a number of ten people they will take you for free round trip.

While the billboards are telling indicators of the landscape of HIV in rural South Africa, I was struck by the limited number. Not counting the signs on display outside of local clinics, there were surprisingly few public markers of the HIV/AIDS epidemic. One must look deeply to see the underlying currents through which disease and health are understood and experienced by local people.

By way of comparison, I drove two members of my research team to Mbabane, the capital of Swaziland. It was a Sunday morning, so it was very quiet on the road and in the local shops. At the border gate, people were dressed in formal clothing as they either walked or (primarily) drove their vehicles across the border. The drive to Mbabane took roughly two hours; we went slowly to examine the landscape. In only one hour of driving, we came across more signs of public health campaigns on HIV than in the entire South African study area. Billboards were on display, telling residents to get tested to know their status or encouraging men to be circumcised. Roughly thirty minutes from the border gate, we stopped at a tourist destination at Pigg's Peak, on a beautiful vista overlooking the valley. Houses and small agricultural fields were spread throughout the surrounding hillside, but, given the late morning heat, there were only a few people out walking. A community education center was easily identifiable by the bright decorations on its exterior. A tree, painted white with black stripes in a zebra pattern, stretched to the ceiling, decorated by multicolored handprints. White doves were painted circling the top of the tree, and the lower branches pointed to two open books that read "Educate the Youth, Educate the Nation." Surrounding the entire structure were painted white candles encircled with the red "AIDS ribbon." It was difficult to pass this structure quickly, to not pause and gaze at the beauty of the painting and feel a sense of hope, a feeling of purpose articulated by members of the community in the face of pressing health challenges.

More public-health billboards were visible on the road to Mbabane. The first had a large fist with the thumb sticking up and read "A Man Knows to Be the Best—He Has to Test." Farther south, in the town of Pigg's Peak, there were multiple posters attached to light poles, with a single figure that appeared to be male, darkened in an outline style. The accompanying message stated: "Choose one partner, get more time, more love, more life." On top of it was a sticker, "Choose one: get it all," with two figures holding hands with each other and two other partners that were crossed off. A Facebook page was highlighted as part of the campaign at facebook.com/khetsamunye. As we returned to the border, we saw yet another that depicted a stylish young man with sunglasses over his fore-head. In SiSwati the sign read "Lisoka Lisoka Ngekusoka," which in English means "I am doing all I can to conquer HIV." As in South Africa, military terms are used to describe the onslaught of the epidemic and the need for aggressive responses to win the battle. HIV is a threat, an invader on the landscape that must be conquered. We left Swaziland impressed by the amount of HIV information on display, compared to what was visible in South Africa. Yet what is specifically being communicated by these signs on the HIV landscape? And what is left unsaid that remains mean-ingful for those living with disease and those vulnerable to health chal-lenges in the future?

BIOMEDICAL HIV

These billboards and other signs are representations of public health cam-paigns in the Swaziland landscape. Like similar messages in South Africa, they emphasize personal responsibility. Individuals are expected to get tested, to be faithful to their partner, and (if male) to be circumcised to reduce the chances of acquiring the disease or passing it on. The body, the individual, the individual decision maker: these are the contours through which public health messages are communicated about HIV. The body itself becomes a landscape on which the epidemic is inscribed. These mes-sages reinforce a behavioral approach to public health whereby the way to reduce the spread of infectious disease is to change behavior—which, in the case of South Africa, largely involves sexual decision making. It is also notable that these depictions are masculinized by positioning men as cen-tral to the public health campaigns against HIV. Sexually active men are depicted as the drivers of the epidemic. In terms of advocating male cir-cumcision, this is not surprising. But the other messages also ensure that men are put on an equal footing with women, if not a higher level of responsibility, in combating the HIV/AIDS epidemic. This has parallels with HIV prevention campaigns elsewhere, such as in Los Angeles, where it has been reported that the male is "made healthy socially by taking on an individual responsibility that is simultaneously a community responsi-bility—the individual subject remains the center of the epidemic and the center of the response to reduced HIV transmission."[21] The South African public health campaigns focus on heterosexual contact while attending to the sexual network that is believed to contribute to the high rates of incidence in these areas. Receiving education about disease causation, being faithful to one's partner, and reducing the size of one's sexual network are all emphasized. These messages are personal, intimate, and individualized. What they fail to do is locate that individual in broader political economies and structural forces that generate disproportionate vulnerabilities to HIV.

The biomedical model serves as the foundation of Western biomedicine and shapes dominant views about human health and disease. As discussed above, the biomedical model has the potential to downplay the role of social factors in shaping exposure to infectious disease and vulnerability to the conditions that produce noninfectious disease. What this means for HIV/AIDS is that biomedical HIV follows a simple, linear, cause-and-effect framework whereby reducing the transmission of HIV is the key to addressing the AIDS crisis. Researchers working in South Africa argue that education programs alone are not the solution to preventing disease transmission, because sexual decision making remains situated within social and gendered norms that increase the vulnerability of certain populations. For example, Mark Hunter has argued that low marriage rates and the coincidence of wealth and poverty in close proximity drive gender relations and sexual relationships that have contributed to high transmission rates in KwaZulu-Natal.[22] As explained in chapter 2, perceptions of HIV and wellness are produced by a variety of factors, such as access to healthy food, gender dynamics, socioeconomic poverty, and views on traditional medicine. The simplification of disease causation to the transmission of the virus from one body to another through sexual contact narrows the scope, moving away from the social and cultural forces that shape individual possibilities and decision making. Interviews with residents in these areas challenge the removal of societal factors from analyzing the spread and response to human disease, suggesting that holistic and contextualized, rather than simple cause-and-effect, perspectives are needed. It is not complete to say that exposure to HIV results in the spread of the disease within a human body; rather, it is necessary to attend to the sociopolitical dimensions that serve to make individuals, families, and communities differentially vulnerable to illness and the opportunities for its management. These vulnerabilities need to be addressed because they speak to the underlying structural determinants of health.

Earlier I referenced an interview with Mafuane, who is a resident of the Mzinti RDP housing project. Mafuane is a friend of Elinah, and they occasionally work together catering events in the area. As Elinah and I drove through the RDP the previous week, she saw us while sitting on her front step and asked if we were going to visit with her sometime. When Elinah had interviewed her the previous year for the structured survey, Mafuane had disclosed her status as HIV-positive. When I spoke with her that day, she shared that she had tested positive three years earlier after having a stroke. While she was not on ARVs at the time, she was taking boosters because her CD4 count was 650. This is an encouraging number and reminded me that the CD4 count came up in nearly all of the interviews with HIV-positive individuals. They are intensely aware of their count and use it to guide decisions about their personal health management. It is a keyword between the patient and the clinic that informs the course of treatment provided, and it is a marker of the embodied experience for each individual living with HIV. Mafuane visits the clinic every six months to get her blood tested. She doesn't take the booster every day but typically once a week when she lacks healthy food or if she is not feeling well at that time. I asked if she received any classes from the clinic, and she said no, but they told her to minimize her stress, condomize if sexually active, and eat healthy foods such as vegetables. Mafuane noted that the clinic also said she should not use traditional medicine because some of it can give you diarrhea and make you ill. When asked when she might go on ARVs, she explained that when her CD4 count fell below 350 she would start treatment.[23] Once that happened, Mafuane anticipated that she would take classes in which she would be counseled about the drugs and how they work. Mafuane emphasized that things have changed for her. Specifically, she said that previously she was not taking good care of herself, but she was now.

We then talked about health and environment and changes due to HIV. Mafuane noted that in the past, people would see the sangomas, but they would cut you and use the same razor with multiple people. In the past, people would get tuberculosis and then die, while now they can go to the clinic for treatment. Also, the sangomas refer people to the clinics to get tested to know their status. When asked if people are willing to share their status, Mafuane explained that people are more open than in the past, noting that previously when someone tested positive they were going to die, but now people see that the infected can live longer. Mafuane credited the ARVs for contributing to this change, sharing that her sister had tested positive for HIV two months ago and was put on ART. She commented how she had made an example of herself for her sister, to encourage her to get tested. For about a week her sister had side effects and was weak, but now she was strong. I was reluctant to ask much about her past, but she began talking about the gendered divisions in the community, noting that it is the women who are willing to get tested and take ARVs. By comparison, men are "stupid" and are less willing to take care of themselves.

The political environmental context of HIV in South Africa results in different lifeways for men and women. Understandings of the virus are embedded in cultural and gender dynamics that shape the ways in which HIV is embodied and managed. Women are differentially vulnerable because of insecure land-tenure systems and power dynamics that influence decision making, which make it difficult to negotiate the terms of sexual contact. Stigmas vary as well. In interviews I conducted in 2014 and 2016, stories were shared of women in the community who were blamed for bringing HIV into the household. In one case, a woman was kicked out of the home and forced to move elsewhere, while her husband walked around with visible signs of the disease. In other conversations I was told that women are more likely to get tested for HIV, and more likely to adhere to ART if they test positive. As Mafuane insisted, women are more responsible in taking care of themselves but then must navigate a health-care system that generates tension with their responsibility to their families. While the billboards on display are likely designed to confront previous instances of women being blamed as the cause of HIV transmission, they also minimize local dynamics that produce the landscapes of HIV.

By locating disease causation at the site of the individual body, the biomedical model contributes to the behavioral model of public health in placing the responsibility on the individual social actor to reduce his or her own vulnerability. The few billboards on display in the South African region, like the others scattered throughout Swaziland, are all directed at individuals, making personal invocations and admonishments toward the man or woman vulnerable to infection. In South Africa, women are more likely to be infected with HIV than men, a fact that is driven by multiple social factors, including unequal sexual decision-making power and intra-household dynamics. As a result, providing mass contraception and absti-nence education will do little to reduce sexually transmitted diseases if women have limited power in negotiating sexual practices. South Africa has been active in promoting condom use to prevent the spread of HIV, but social relationships that intertwine with sexual decision making create an uneven landscape. Additionally, as with human health itself, public health campaigns can be fleeting. I returned to the study area in May 2014 to undertake interviews with residents who had participated in the struc-tured survey from the previous year. A circumcision clinic had opened in Mzinti during this time but had promptly closed because its funding ended. I also noticed that all the billboards had been taken down, with nothing put in their place. In fact, I could find no public-health billboards in the area. Their skeletons stood vacant, marking the end of another pub-lic health campaign.

Finally, a behavioral emphasis in public health campaigns against HIV directs critical resources toward prevention as opposed to treatment, thereby potentially limiting the response for those already infected. Given that the home-based care groups in the region are composed primarily of women, who are often volunteering their time rather than receiving a con-sistent salary or adequate resources, this is particularly glaring. The front lines of HIV care services, referred to in the opening of the chapter, are being provided by women. In an early assessment of the global response to HIV/AIDS, Peris Jones argued that the Western donor community was "largely silent" about ARVs and more insistent on preventing the spread of the virus.[24] Biomedical HIV, therefore, has a particular focus on limit-ing the transmission of the virus at the level of the individual body by inducing behavioral change that involves personal responsibility to stop the epidemic. This narrows the potential pathways for how HIV is under-stood within socially and spatially dynamic communities like those in South Africa. It also filters perspectives of the landscapes of HIV so that those in full view are the individual bodies infected or those made vulner-able to infection. The underlying political environmental context recedes to the background, even though it fundamentally contributes to the creation of these vulnerabilities. The production of HIV landscapes is ongoing, however, and they continue to change as a result of the increased availability of ARVs that have made HIV lifeways a possibility for many.

THE POLITICAL ENVIRONMENTAL CONTEXT OF MANAGED HIV

The construction of colonial and apartheid spaces of segregation continues to have lasting material impacts on the social and ecological systems that shape the political environmental context for managed HIV. Vulnerabilities to HIV exist through social relationships and power dynamics that are produced through spatial processes. The result is a territorialization of disease generated through disproportionate vulnerabilities and unequal opportunities for health management. The landscapes of HIV in the contemporary period, therefore, are tangible reminders of spatial configurations that have been created and contested over centuries. The possibilities for HIV lifeways exist through the capacity of individuals to negotiate a range of services, including infrastructure deficits that create unequal health-care services in rural areas. High population densities that have resulted from apartheid spatial policies, coupled with land tenure systems that preclude opportunities for agricultural production, generate food insecurities that are intensified by adherence to ART. Access regimes continue to be mediated by a mix of institutional arrangements and can limit resource-collection practices that contribute to livelihood possibilities. These conditions are what make South Africa's HIV/AIDS epidemic, and the current experiences of those managing HIV, a distinct state of disease.

The national government recognizes differential vulnerabilities to infectious disease among its population. The National Strategic Plan for 2012–16 emphasizes multiple social factors that produce higher vulnerabilities to HIV transmission.[25] First, there is a notable recognition of the gendered dimensions of HIV. The plan indicates that young women aged fifteen to twenty-four are four times more likely to have HIV than men in the same age cohort and are likely to become HIV-positive five years earlier than men.[26] Secondary schooling is identified as "protective against HIV, especially for young girls." Uncircumcised men, transgendered populations, and homosexual men are listed as having a higher risk of acquiring HIV. Orphans and other vulnerable children and youths are also identified as an at-risk group. Second, those living in poverty are more likely to have HIV. This is highlighted by some of the factors listed as producing HIV vulnerabilities, such as the use of illegal substances, alcohol abuse, and sex work. Third, socially produced infrastructure, such as proximity to national roads and highways, in addition to informal settlements in urban areas, are identified as factors that contribute to HIV transmission. Lastly, migration is mentioned as increasing the risk of acquiring HIV. The Global AIDS Response Progress Report from 2012 concludes its discussion on HIV vulnerabilities by noting that

> a lack of adequate services, in addition to several social and structural barriers that include stigma and discrimination, have significantly contributed to the disproportionate HIV prevalence present among key populations in South Africa. By investing in the specific sexual and reproductive health needs of key populations at increased risk of HIV acquisition, the number of new infections would be reduced enormously.[27]

While reliable data on HIV prevalence in the country are incomplete and sometimes inconsistent, a review of some of the existing sources shows a disproportionate epidemic that overlaps with historical spatial configurations. A 2005 report from the Human Science Research Council noted that nearly four times as many women as men aged twenty to twenty-four were HIV-positive (23.9 percent vs. 6 percent) and that many of these women had older partners.[28] The 2008 South African National Survey also showed variation by population group, with those classified as African having a prevalence of 13 percent, compared to 1.7 percent for "coloured" and less than 1 percent for those classified as white or Indian.[29] The four provinces that had the highest estimated rates of prevalence in 2008 were KwaZulu-Natal, Mpumalanga, Free State, and Northwest. These provinces contain all or part of six of the ten Bantustans, specifically KaNgwane, QwaQwa, KwaZulu, KwaNdebele, Bophuthatswana, and the southern section of Gazankulu. According to an antenatal seroprevalence survey in 2011, the highest rates of prevalence were 39.5 percent in KwaZulu-Natal, 35.1 percent in Mpumalanga, 30.6 percent in Free State, and 30.4 percent in

Gauteng.[30] As with the 2008 National HIV Survey, these provinces contain territories that were part of the Bantustan system. Looking within the provinces, a study that applied spatial models and Bayesian prediction to data from a national HIV household survey conducted in 2003 found that the highest rates of HIV prevalence were in northwestern KwaZulu-Natal, southern Mpumalanga, and eastern Free State.[31] While it should be noted that these are predictive models, those areas roughly overlap the former Bantustan territories of KwaZulu, KaNgwane, and KwaNdebele.

In addition to corresponding with higher HIV prevalence than in the general population, historical spatial systems simultaneously mediate the possibilities for accessing health care in the contemporary era. For example, economic disinvestment in the Bantustan territories was justified by asserting their need to develop independently; however, this generated a particular social infrastructure that was inequitable compared to the rest of the country. This has been addressed by the ANC national government over the past twenty years, but change remains slow. Even with the investment in health-care facilities such as the Tonga Hospital, there remains a shortage of capable health-care practitioners to provide much-needed services to rural populations. As noted by Victoria Scrubb, because of limited governmental regulation and little oversight of privatized health services during apartheid, health care in the Bantustans "frequently ignored quality-of-care guidelines and became places of abuse and maltreatment."[32] Scrubb also suggested that the apartheid government limited the budgets for rural health systems, particularly in comparison to urban centers, thereby leaving these areas without needed resources. This has contributed to two tracks in the existing national health-care system: private facilities are available to some, while the public system remains dependent on often underfunded, nurse-run primary-care clinics; district hospitals with generalist physicians; and a diffuse network of regional and tertiary hospitals with specialists.[33] As a result, poorer rural residents continue to face barriers in terms of the distance to facilities, costs associated with transportation, and inadequate services.[34] It has been estimated that poorer households may spend over 10 percent of their household income on health-related transport costs.[35] This is supported by some of my interviews, which reveal that patients utilize a mix of health-care facilities across a spatially diffuse area.

While comparatively good health-care facilities exist in the study area, there are geographic disparities that are recognized by residents. The delay in the establishment of the Ntunda clinic is one clear example. As one member of the men's focus group explained:

> Coming to the issue of the clinic, this is the biggest concern in our community. It is affecting us all. People get sick. They need to go to the clinic. Just because of that patients end up dying. Because of that in case of emergencies the clinic that you have to go to is far from your village. Sometimes you will find people who are staying home, some want to check their health status. But they cannot because the clinic is far from them.

Even in cases where residents were able to travel to the Mzinti clinic, it was explained that the waiting list could be quite long, so it could take much of the day to receive care. One participant in the Ntunda men's focus group complained that when he visited the hospital, but especially the clinic, he would be told there were no pills. He accused the nurses of taking the pills and giving them to their relatives, because the nurses are not searched by the security guards at the gate. Even though the government provides the pills, he complained that the nurses "end up cutting medicine and adding water out of bottles." These types of statements attest to the multiple perspectives about the sites of care for local residents, which are mediations between the state and its citizens. These encounters underscore the complex links between health care and poverty, as well as underlying spatial formations that shape accessibility to health-care services.

The inequities in provision of services and economic development have not gone unnoticed by community members more than twenty years after the democratic transition. Younger residents complain about the lack of opportunities being provided by the government and increasingly question whether the ANC is still committed to social and economic change. As one participant in the Schoemansdal focus group poignantly detailed:

> The way we live here there are no services. What we see is only water and electricity. We all live for government grants. . . . Even where we stay there is no road to the graveyard so we have to take the coffin and hold it with our hands until we reach the graveyard.

In many interviews, the state of the hospitals was described in terms of the demand for care and lack of needed services. Community members noted the long lines to wait for care and also the absence of medicine at certain times. The consequence is that human health is not simply the product of disease transmission but the way in which the health domain is enabled and constrained by the political environmental context. Simply put, people in Mzinti, Ntunda, and Schoemansdal are not unwell solely because of HIV. In the era of managed HIV, people are unwell because of their inability or unwillingness to seek out care and medical treatment. The constraints to managing their health are not simply the product of contemporary dynamics but also result from deeply rooted inequities that have been spatially produced.

The construction of core–periphery dependencies between cities and urban townships, and between white-owned rural areas and the Bantustans, produced migration patterns that reduced longer-term investment in the Bantustan territories. While the apartheid state asserted that these were independent nations in the making, the Bantustans were deprived of the material resources, both human capacity and capital, that would have facilitated economic growth and development. Agricultural production was emphasized as central to rural livelihoods, even as prime land was removed from African control. The former Bantustans have high population densities compared to other rural areas in Africa, which has continued impacts on livelihood possibilities, including agricultural production. Population densities in rural South Africa are notoriously difficult to estimate, but as one example, Marshall Murphree (1990) reported a human population density in the communal areas of Zimbabwe of five to ten people per square kilometer. By comparison, the human population density in the communal areas of the Mpumalanga and Limpopo provinces in the same year was calculated as 174 people per square kilometer.[36] When there has been state investment in agriculture, it has not been directed to household consumption but to international export, as evidenced by the presence of sugarcane farming in the region. The global demand for biofuels only intensifies the investment pressure for export commodities, as opposed to foods needed locally for daily consumption. Driving into areas that constituted the Bantustans during apartheid—the same areas that have higher estimates of HIV infection than the national average—is especially striking because it is

necessary to pass vast expanses of privately owned sugarcane farms surrounding the edges of these territories.

Changing environmental conditions can intensify the challenges for
managed HIV. I conducted interviews in Schoemansdal in January 2016 at
the height of a drought that has gripped the southern African region.
Driven by El Niño conditions that brought dry and warm patterns, 2015
was the driest year in South Africa since official records began in 1904. The
South African Weather Service reported that during a heat wave in early
January 2016, thirty-one locations reached new maximum-temperature
records.[37] Food prices are expected to increase, and millions of tons of
maize have been imported into the country. In one interview with an HIV-
positive grandmother named Mandla, she pointed to the garden next to
the household where she grew vegetables for the family. Because of the lack
of rainfall, Mandla noted that she had not planted yet, though she was
hopeful she would soon be able to. As she explained, "We are suffering
because of the scarcity of rain. If there is rain we are able to do it all, we can
grow some sweet potato and cassava, but if there is no water, we are not
well at all." For many of my interviewees, the challenges in producing their
own food from household gardens were compounded by shifting environmental patterns that increased their vulnerability to food insecurity.

Food security is a critical challenge, especially for those adhering to
treatment. Recent studies estimate that those infected with HIV typically
require up to 15 percent more energy and 50 percent more protein, as well
as more micronutrients.[38] The limited options for farming, due to historical systems of segregation, result in pressure on communal areas for
resource collection. Wild foods that are collected include distinct species
such as *Amaranthus* sp., *Chenopodium album,* and *Bidens pilosa* that are
high in carotenoids and vitamin A, nutrients that have been shown to contribute to reducing infection risk and slowing the progression of HIV into
AIDS.[39] Food security involves more than agrarian production on farms or
in gardens next to the household; rather, it extends to the communal areas,
where resource extraction is a safety net for many in the region. Biological
diversity in rural South Africa has been compromised because of high population densities, thereby reducing the availability of flora and fauna for
poor households. In one of the few studies on this topic, Dylan McGarry
and Charles Shackleton identified disease as a trigger to greater economic

vulnerability, which then threatens the viability of ecosystem function-ing.[40] Focusing on the consumption of wild meat in response to AIDS mor-tality in the household, they found that seafood, riverine fish, forest mam-mals, birds, reptiles, and insects contribute in significant ways to children's diets. Highly vulnerable families were observed to hunt more regularly and consume more wild meat than less vulnerable families, which suggests that greater resource pressures might accompany the spread of infectious dis-ease and mortality.[41] Similarly, Lori Hunter and colleagues evaluated lon-gitudinal data from Mpumalanga Province and concluded that adult mor-tality had increased household reliance on natural resources collected from communal areas. Accompanying this change was an increased dependence on collected or grown food for items previously purchased, which, taken together with the loss of the adult household member, contributed to decreased food security for the remaining family members.[42]

Natural-resource needs in rural communities place pressure on the conversion of territory for other development purposes. These patterns also reveal conflicting institutional arrangements that determine access to resources critical to livelihood production. While these communal areas were considered state land under apartheid, they were under the jurisdic-tion of the tribal authorities, which included control over the collection of various resources routinely utilized for livelihood production. In an inter-view I conducted in 2001, the induna emphasized the role of the tribal authority in managing these territories for community members. The in-duna noted that

> now we are trying to stop people from cutting trees but they don't listen. You know those people across the Komati River at Mangweni village. They are hardly disturbing us with our trees and they have finished all the trees. . . . Those who want to cut trees and make houses, they must have a note. Once they come to me I take them to the chief to get permission so that they must cut trees. But some people, they don't want to do that. They only go and do that on their own.

Continuing, he emphasized the importance of livestock for community members and the need to maintain spaces for grazing.

The historical spaces of the native reserves and Bantustans were socially engineered for centuries with the express purpose of extracting labor and resources for the benefit of the minority ruling class. Race and class collided in producing geographies that had material implications for human health and well-being. These historical landscapes continue to produce the possibilities for those managing HIV and other illnesses. Whether it is land dispossession that reduced agrarian opportunities that would improve the prospects for food security, absence of economic employment that would generate capital to offset the costs of illness, varying proximities to clinics and other health-care facilities, or entrenched cultural practices that disempower women and other groups, these all have tangible effects on the possibilities for human health and well-being. Sexual decision making is one of the most intimate types of social interaction. South African agencies have encouraged people to "condomize" as a way of managing the spread of the virus; however, the use of condoms filters through social, cultural, and gendered relationships that are historical and power laden. The contemporary HIV/AIDS epidemic is a product of historical spaces and political economies that have made some populations differentially vulnerable. Women, those living in socioeconomic poverty, and specific racial and ethnic groups all display variations in the prevalence of HIV compared to the national population. The Bantustans were socially constructed territories that continue to produce the possibilities for human health and well-being. This is the political environmental context that creates the landscapes and lifeways of HIV.

5 Health Ecologies within Dynamic Systems

Human health exists at the nexus of social and ecological systems and is shaped by their coupled and dynamic interactions. While socioeconomic development often provides access to health services, its broader patterns transform landscapes and generate exposure to conditions that can contribute to poor health. Ecological systems similarly play a formative role in determining how various functions, such as water discharge, biodiversity, and soil fertility, produce human health. Precipitation patterns and other weather systems can shift habitats for disease vectors, such as mosquitoes, or reduce the quality of agricultural production, thereby making people more vulnerable to nutritional deficits and reduced caloric inputs. Interactions between pathogens and keystone species, a critical indicator of ecosystem health, disrupt landscape patterns for humans and other species, putting them into contact in ways that influence disease transmission.[1] Social and ecological systems do not operate independently but in concert. Situated at these points of intersection is the political environmental context, which generates the conditions that make people vulnerable to poor health. The political environmental context can be quite dynamic, given transformations in social and ecological systems, thereby resulting in fluid health domains for human populations.

The previous chapters examined the landscapes and lifeways of HIV in South Africa to demonstrate how disease is embodied, experienced, and managed. Managed HIV is produced by various social dynamics that influence access to ARVs and other public health services. Popular understandings of HIV/AIDS are interlinked with previous decades of external intervention and racial classification that informed attempts by the state to respond in particular ways regarding the provision of ART and other types of public health interventions. Vulnerabilities to HIV infection are produced through the complex interplay of race, gender, poverty, historical spatial economies, and disease discourses that result in uneven lifeways and disproportionate challenges for HIV management. State policies contribute to the discursive context in which diseases are understood, producing disparities in responses to transmission and treatment. As discussed in chapter 4, biomedical HIV focuses on the individual body as the site of HIV management. Disease transmission is primarily understood as occurring through contact within sexual networks that increase exposure to the virus. Representations of biomedical HIV articulate messages attached to gender dynamics in rural areas that minimize the structural context that produces vulnerability. Biomedical HIV places responsibility on those who experience greater vulnerabilities due to underlying spatial dynamics. While spatial processes are fundamental to the lifeways and landscapes of HIV, the natural environment is also critical in facilitating access to nutrition, resisting competing infectious agents that increase vulnerability, and supporting agricultural production for food security. For those infected and affected, the daily management of HIV intersects with livelihood possibilities generated by environmental conditions that have been produced by historical spatial economies.

My intention in this chapter is to extend the analysis beyond managed HIV in South Africa to examine how social and ecological systems produce dynamic health domains for populations in other settings. Specifically, I address the ways in which environmental dynamism reshapes the political environmental context for human and nonhuman agents. In order to do so, I draw upon the field of environmental health to consider the interconnected and dynamic flows between human and nonhuman nature that produce the conditions shaping health and well-being. The natural and built environment can contribute to the spread of disease or to the possibilities for healthy decision making. These dynamics necessitate a historical and spatial approach that is sensitive to the complex interrelationships between human and nonhuman species, and in natural and physical settings. In particular, global climate change will challenge the future dimensions of environmental health. Therefore, I consider how weather patterns, environmental variability, and other aspects of climate change influence vulnerability to infectious disease while undermining the capacity of human systems to adequately respond. These ecological patterns are simultaneously mitigated and exacerbated by existing sociopolitical structures that shape the political environmental context for human health.

I follow this engagement with environmental health and climate change by examining how shifting environmental patterns in northern Botswana are affecting social and ecological systems in the region. Beginning in 2008, increasing floodwaters in the Okavango Delta have displaced human settlements in the secondary floodplain, such as in the Etsha region on the western edge of the delta.[2] The increasing volume of water in the Okavango Delta has similarly fed tributaries running to the south, including the Boteti River. Though dry for two decades, the return of water in the Boteti riverbed is reworking the interactions between human livelihood systems and ecological processes. The newly arrived water is presenting expanded livelihood opportunities that might improve health into the future but is also perceived by some residents as generating vulnerabilities to infectious diseases such as malaria and lumpy skin disease. Drawing from fieldwork in the Boteti region, I emphasize the interactions between these ecological shifts and existing social structures, as well as the larger context that dictates responsiveness to these changes, as the point of contact between people and disease patterns that converge in distinct places and landscapes. Disease vectors are simultaneously political and ecological. Their production demonstrates the role of the political environmental context in producing human health. While water is the lifeblood of the region, its return is complicated and feeds into preexisting political and economic systems that result in differential vulnerabilities to infectious disease and uneven capacities to manage health and well-being.

ENVIRONMENTAL HEALTH AND CLIMATE CHANGE

Environmental health is a field within public health that addresses the ways in which the natural, physical, or human-constructed environment shapes human health and well-being. Environmental health concerns the built landscape, exposure to environmental toxins, and access to green space. Various international agencies recognize the role of the natural environment in shaping disease occurrence and outline recommendations to generate opportunities for improved health. The WHO notes that millions of deaths could be prevented each year by addressing environmental hazards such as unsafe drinking water or exposure to air pollution.[3] In developing countries, the main environmentally caused diseases are diarrheal disease, lower respiratory infections, unintentional injuries and accidents, and malaria. Physical injuries are responsible for more than five million deaths each year, roughly equal to the number of deaths from HIV/AIDS, malaria, and tuberculosis combined.[4] Low- and middle-income countries account for 91 percent of deaths due to unintentional injuries.[5] In wealthier and industrialized countries, cancer, cardiovascular disease, asthma, lower respiratory infections, and traffic injuries are among the primary environmental health hazards. According to the WHO, 85 of the 102 categories of diseases and injuries listed in the *World Health Report* are influenced by environmental factors. Healthy and sustainable environments are needed to reduce morbidity and mortality in human populations. It is therefore necessary to identify future challenges that might be generated by environmental change, such as those caused by extreme weather events, or variability in the amounts or timing of precipitation, resulting from global climate change.

According to the "WHO Public Health & Environment Global Strategy Overview," environmental hazards influence over 80 percent of communicable and noncommunicable diseases and injuries.[6] Overall, environmental hazards are responsible for roughly a quarter of the total burden of global disease. In developing countries, the burden of environmental hazards is more strongly experienced by communicable diseases such as malaria, Dengue fever, and HIV. Environmental health is not strictly focused on infectious diseases but also on noncommunicable diseases, which the United Nations has recently identified as the primary cause of death and disability in the world. Roughly two-thirds of annual deaths are caused by cardiovascular diseases, cancers, chronic respiratory diseases, and diabetes. Because many of these are thought to be preventable, public health experts emphasize the importance of reducing tobacco and alcohol use, unhealthy diets, and lack of exercise.

The challenges for maintaining sustainable environmental health are compounded by a global population that is becoming increasingly urban. According to WHO estimates, 70 percent of the world's population will be living in towns and cities by 2050.[7] While this can mean greater access to health-care services and amenities compared to rural areas, urbanization can intensify health risks. Globally, road-traffic injuries are the ninth leading cause of death, and most road-traffic deaths occur in low- and middle-income countries. Urban air pollution kills around 1.2 million people globally each year, primarily from cardiovascular and respiratory diseases. A significant proportion of urban air pollution is caused by motor vehicles, although industrial pollution, electricity generation, and household fuel combustion in resource-poor settings are also major sources. Lastly, the WHO notes that urban environments are not always designed to support physical activity or promote healthy food consumption. The opportunities for physical activity can be constrained by a variety of urban factors, including overcrowding, high-volume traffic, heavy use of motorized transportation, poor air quality, and lack of safe public spaces and recreation facilities. Thus, urban environments can generate poor health outcomes for human populations, though within these settings such patterns are not equally experienced. As the earlier discussion on environmental justice shows, the places and landscapes of disease have been produced through historical spatial processes that result in inequitable exposure to carcinogens and other factors that decrease health possibilities for some members of the population.

Anthropogenic climate change represents an existential challenge for environmental health. Variations in the amount, intensity, and duration of precipitation can alter the environmental conditions that allow certain disease vectors to survive.[8] Additionally, temperature changes can shift the latitudes for ecosystems and allow species to move into new areas.[9] In order to explore some of the social and ecological connections to human health, I now turn to research reviews from the IPCC, specifically its recently released Fifth Assessment Report. The IPCC was created by the United Nations Environment Programme and the World Meteorological Organization in 1988 to provide the global community with the most current and reliable scientific information about climate change and its potential impacts on environmental and social systems. Approved by the United Nations General Assembly, it reviews and assesses the most recent scientific, technical, and socioeconomic information on climate change. This information is provided in reports released to the general public, of which the Fifth Assessment Report, disseminated by three working groups in 2013 and 2014, is the most recent. The report includes a chapter titled "Human Health: Impacts, Adaptation, and Co-benefits" that addresses the links between global climate change and human health, in which three distinct pathways are identified.[10] The first pathway includes direct impacts of climate change that are tied to shifts in the frequency of extreme weather events, including heat, drought, and heavy rain. Human populations could be disrupted by rising sea levels, reduced access to environmental services, and agricultural decline.[11] The second pathway includes the effects mediated through natural systems, such as the roles of disease vectors, waterborne diseases, and air pollution. The IPCC indicates that greater risk of injury, disease, and death will result from more intense heat waves and fires, noting with high confidence that there will be increased risk of undernutrition stemming from reductions in food productivity and increased risks of foodborne, waterborne, and vector-borne diseases. The third pathway addresses the effects of human systems, such as occupational impacts, undernutrition, and mental stress.

The IPCC report emphasizes that the coupled links between climate change and disease vulnerabilities are still emerging and necessitate urgent research attention. Additionally, there remain striking variations in vulnerabilities to climate change in terms of human health and well-being. The report notes that until mid-century, climate change will most likely intensify existing health challenges in ways that affect the most vulnerable. The most vulnerable populations are anticipated to have reduced work capacity and less productivity, which could cause a downward spiral into increased socioeconomic poverty. In areas with high HIV prevalence, climate extremes will tend to multiply health risks because of the interactions between poor health, poverty, and undernutrition.[12] In South Africa, this presents new challenges for managed HIV, given the importance of food security for those on ART. In many interviews, residents shared the challenges they face in generating high-quality food to meet their nutritional needs while taking ARVs. In one lengthy interview conducted in 2014, a resident of Mzinti talked about the small garden that she was growing next to her main compound. She had been advised by the clinic to grow vegetables, but she was struggling to maintain the plot because of a lack of water—a challenge that has likely been intensified by the pronounced drought of recent years. Shifting weather patterns associated with global climate change are anticipated to complicate the management of HIV in numerous ways, including the relationships between livelihood systems, agricultural production, and food security.

The IPCC report points out that until the middle of the twenty-first century, climate change will largely intensify already existing health problems, with the largest risks to populations that are most directly affected by climate-related diseases. This means that vulnerabilities to weather patterns and other aspects of climate change that influence human health vary geographically. Poorer regions with populations that are more dependent on agricultural production for subsistence are less likely to have the resources needed to adapt to change. Similarly, countries that currently lack adequate health-care facilities will be more threatened by new infectious-disease patterns caused by shifting ecosystem dynamics. There are also conditions within populations that generate differential vulnerabilities to climate-related injury and illness. For example, the negative effects of malaria, diarrhea, and undernutrition are more widely experienced by children.[13]

Because of the interconnections between ecological dynamism and social systems, the IPCC and other institutions highlight the need to reduce vulnerabilities by increasing access to clean water and sanitation; improving health-care services, including vaccination and pediatric care; reducing socioeconomic poverty; and improving the ability of populations to respond to disasters. Addressing vulnerabilities also requires attention to a global population that is urbanizing. The urban heat-island effect is a leading cause of weather-related mortality in the United States and is expected to intensify elsewhere as a result of the growth of urban populations and rising temperatures.[14] The young and elderly are more vulnerable, as are those without access to air conditioning. It is generally anticipated that environmental changes, such as flooding and agricultural decline, will result in the displacement of populations and potentially contribute to social conflict over scarce resources and territory. At the 2012 meeting of the American Association for the Advancement of Science, it was announced that the United Nations has projected there will be fifty million environmental refugees by 2020.[15]

POLITICAL ENVIRONMENTAL VECTORS
AND MALARIA (RE)EMERGENT

Changing weather and climatic effects are anticipated to have consequences for the spread of infectious diseases to human populations, although establishing direct links is complicated by a mix of social, economic, and environmental factors. Global climate change is transforming ecosystems; even small increases in temperature can facilitate the spread of disease vectors into new geographic areas. Disease vectors, the most studied of which are mosquitoes and ticks, are organisms that carry infectious disease and transmit it to other organisms. For example, the mosquito *Anopheles* carries a parasitic protozoan in the genus *Plasmodium*. More than a hundred species in this genus cause infection in reptiles, birds, and certain mammals.[16] Currently, four types of *Plasmodium* infect human beings, specifically *P. malariae, P. ovale, P. vivax,* and *P. falciparum.* A fifth type, *P. knowlesi,* naturally infects macaques and has recently been recognized as a cause of zoonotic malaria in humans. There were an estimated 216 million episodes of malaria worldwide in 2010, primarily among children under five years of age in Africa.[17] *Anopheles* carries the malaria parasite and infects humans by transmitting it to an individual's blood. Symptoms, including fever and chills, typically occur within a couple of weeks and can lead to death. The effectiveness of mosquito-to-human transmission is tied to social and ecological conditions that provide suitable habitat for the mosquito to breed in sufficient numbers to make transmission more likely. In a historical assessment of malaria patterns, Randall Packard explained that ecological changes associated with the rise of stable agricultural populations in Africa contributed to the transmission of malaria in humans in one more important way. They played a part in the evolution of the *Anopheles gambiae* mosquito, the most efficient transmitter of malaria in Africa, if not the world.[18]

Anopheles mosquitoes are highly dependent on the natural environment, particularly in terms of precipitation and temperature, which can alter the environmental conditions that allow for their survival. One setting that has received international attention for the reemergence of malaria is the Kenyan Highlands. The highlands are typified by a high altitudinal climate zone with shifting ecological conditions along the gradient moving up the mountains. Because of its topography and altitude, it represents a distinct ecosystem that had remained relatively stable but has experienced change in recent decades. In a report in the *Lancet*, Zoe Alsop noted that malaria was eradicated in the 1960s as a result of cooler temperatures and the use of insecticides.[19] But rising temperatures allowed the disease to spread to communities such as Kisii during the late 1980s. There is not a consensus in the scientific community that temperature change is the direct cause of the reemergence of malaria in the highlands, given the compounding factors that shape disease transmission. One such factor, the influence of temperature on malaria, appears to be nonlinear and is vector specific.[20] Recent work has revealed that daily fluctuations in temperature affect parasite infection, the rate of parasite development, and components of mosquito biology that combine to determine the intensity of malaria transmission.[21]

Establishing direct correlation between climate change and disease patterns is a challenge, especially given the multiple ways in which climate change operates. As evidence of this, in the journal *Emerging Infectious Diseases,* Dennis Shanks and colleagues asserted that increased drug resistance is likely the cause of reemergence in the Kericho district in the Kenyan highlands, rather than climate change.[22] Similarly, the IPCC Fifth Assessment Report indicates that determining the role of increased local warming (confirmed by aggregated meteorological data) in the period since 1979 is constrained by the lack of corresponding time-series data on the levels of drug resistance and intensity of vector-control programs.[23] The report explains that "decadal temperature changes have played a role in the changing malaria incidence in East Africa. But malaria is very sensitive also to socioeconomic factors and health interventions, and the generally more conducive climate conditions have been offset by more effective disease control activities."[24] The consequence of these competing views is that human and nonhuman species are directly implicated in the reestablishment of malaria in geographic areas where it was no longer evident. Additionally, the relationships between vectors and people are dynamic and can shift because of variations in temperature, precipitation, or ecosystem functioning. The political environmental context for malaria is therefore produced by social and ecological dynamics that are unfolding over time and space.

Regardless of the difficulties in drawing concrete links between weather-related patterns and infectious disease, transmission dynamics are anticipated to shift into new areas in future years. Yet the spread of these vectors, whether mosquitoes or other nonhuman species, is facilitated through existing social conditions that enable transmission while constraining effective public health responses. Transmission of a particular virus is not equally experienced among individuals within societies and can be driven by underlying political and economic conditions that make people differentially vulnerable to exposure. Vulnerabilities to disease are not equally experienced across a changing world, but rather filter through existing conditions that are unequal and power laden. Given recent critiques of emergent infectious diseases, I emphasize that I am using the concept of reemergence to refer simply to the presence or prevalence of a disease in a location where it had existed previously. In particular, I am mindful of the note of caution raised by Paul Farmer, who suggested that the concept of emergence has "symbolic burdens." As he explained:

> If certain populations have long been afflicted by these disorders, why are the diseases considered "new" or "emerging"? Is it simply because they have come to affect more visible—read, more "valuable"—persons? This would seem to be an obvious question from the perspective of the Haitian or African poor.[25]

Earlier in the book, I argued that biomedical HIV represents a particular understanding of HIV transmission and management. In concentrating on the virus while advocating behavioral responses to minimize its transmission, biomedical HIV narrows the scope from the underlying structural conditions that produce different vulnerability to transmission and opportunities for managed HIV. Much like HIV interventions, anti-malaria campaigns are imbued with material and symbolic meanings that shape perceptions of the disease and the ways in which responsibility toward its prevention is understood. Malaria campaigns often emphasize the causes of transmission while promoting behavioral responses to mitigate its spread. Particularly in Sub-Saharan Africa, there is a focus on stopping transmission and improving treatment for those who are infected. In this case, insecticides and bed nets are routinely identified as the best strategy for eradication, even though recent studies indicate that public health campaigns might be more effective if they followed strategies employed in the southern United States in the early twentieth century.[26] Specifically, large-scale drainage projects that eliminated breeding grounds have been identified as the cause of the elimination of malaria in the United States.[27]

Malaria control is therefore a discourse that varies depending on the setting in which it is employed. Much like HIV, it is differentially experienced—not only because of geographic region or location, but also as a result of social and ecological conditions that are contextual and relational. The discourse of malaria is powerful and can be utilized by national governments to exert control over territory and populations. An example of this comes from Eric Carter's historical analysis of malaria in Argentina, which shows the ways in which a particular disease is bound up with much more than its etiology, including the nation's identity, economic development interventions, or conflicting scientific paradigms.[28] Carter detailed how the discovery of the disease in northwestern Argentina in 1890 generated public health campaigns that correlated the eradication of the disease with the well-being of the nation. This produced development interventions, including agricultural intensification and urban sanitation, that often proved unsuccessful in eradicating the disease. As Carter noted, "Taken together, concern over regional development, anxiety over the fate of the Argentine race, and the etiological framework of miasmatic theory and medical geography buttressed the claims of provincial elites and public health advocates who promoted malaria control as a pathway to social and economic progress."[29] This statement resonates with my earlier discussion of biomedical development. By coupling the management of malaria with socioeconomic development, the Argentinian government generated a teleology in which the eradication of the disease was likened to the health of the body politic. The state of the nation was dependent on controlling malaria, but a close reading of the history of the campaigns reveals specific agendas and power dynamics. Additionally, because malaria has been eradicated in certain regions, particular understandings arise of what it represents and how it can be managed. As Packard has shown, the status of malaria as a "tropical disease" comes after centuries of active intervention by state agencies to manage social and ecological landscapes.[30] The naming of malaria as a tropical illness underscores the importance of discourse within the political environmental context of human health. Casting a particular disease in such terms ensures specific understandings that can inform resulting policy interventions. As Meredeth Turshen argued:

> Many diseases—like malaria and yellow fever—now called tropical were found in the temperate climates of Europe and North America well into the twentieth century. The label 'tropical' reinforces the impression that natural conditions like climate rather than economic conditions or political circumstances are responsible for the persistence of these diseases in the Third World.[31]

Similarly, Packard noted that

> the spread of malaria into Europe and the Americas was closely tied to the changing fortunes of agriculture. In some areas, such as on the island of Sardinia, or the early agricultural colony of South Carolina, or the upper Mississippi Valley, it was the initial opening up of the land for cultivation that transformed the region's ecology and created opportunities for the expansion of malaria.[32]

The transformation of agricultural landscapes in the continental United States over time has played a direct role in making people vulnerable to the disease. This demonstrates that health possibilities are often coproduced through the interactions between humans and the natural environment. As Packard explained:

> In order to understand and to respond effectively to the persistence of malaria as a global health problem, it is critical to view the disease as part of

a wider historical narrative in which human actions have encouraged the breeding of malaria vectors, exposed populations to infection, and facilitated the movement of malaria parasites.[33]

While the trajectories of global climate change and environmental health are uncertain, it is generally accepted that there will be disruptions to existing dynamics in the future. The natural environment has a direct impact on human health, whether in terms of establishing the conditions that contribute to noncommunicable illness or in creating encounters between human and nonhuman species to transmit infectious disease. Changing environmental conditions necessitate attention to how they influence the possibilities for human health and well-being. As the remainder of this chapter details, variability in flooding dynamics in the Okavango Delta of Botswana has transformed the lived experiences of residents in the region. This is particularly the case along the Boteti River, where water has returned to a riverbed that had been dry for decades. The return of the water has meant shifting health ecologies for individuals and families living along the water's edge, in addition to nonhuman species such as livestock that are threatened by vectors that could compromise their health. Dynamic health ecologies along the Boteti demonstrate that human health and well-being are intimately connected with the flow of water in ways that reveal disparities in the political environmental context of the region.

SHIFTING HEALTH ECOLOGIES ALONG THE BOTETI

Human health vulnerabilities and perceptions of disease are interlinked with short and long-term variabilities in ecological conditions. The Fourth Assessment Report of the IPCC outlined current and future impacts of climate change, emphasizing the likelihood of increasing variability in numerous biophysical processes, specifically temperature, precipitation, and flooding.[34] The Fifth Assessment Report confirmed these findings, with added confidence in the likelihood of climate-driven disruptions to human and environment systems.[35] The anticipated impacts from climate change will not be equally experienced. Low-lying coastal communities and small island nations will be affected by storm surges, coastal flooding, and sea-level rise. Developing economies that are more reliant on subsistence farming face greater exposure to weather-related changes, such as variability in precipitation or drought. The implications for food security and human health are likely significant, and the IPCC emphasizes that these dynamics will "differ substantially depending on baseline epidemiologic profiles, reflecting the level of development and access to clean and plentiful water, food, and adequate sanitation and healthcare resources."[36] Additionally, the impacts of climate change will "differ within and between regions, depending on the adaptive capacity of public health and medical services and key infrastructure that ensures access to clean food and water."[37] These patterns have resulted in the increased coupling of climate change with the concept of justice, because many of the countries believed to be the most vulnerable to environmental variability are also the ones least culpable for anthropogenic climate change.

Working Group II of the IPCC has projected that southern Africa is one of several geographic areas that are particularly vulnerable to global climate change, and that social and ecological systems will likely be disrupted and transformed in future decades.[38] Given the reliance on agricultural production and livestock rearing in southern Africa, even slight changes in precipitation can have serious consequences for food security. It is important to emphasize that the predictive models for southern Africa are not uniform in their assessments of these changes, with some studies predicting wetter seasons while others anticipate drier months than are currently experienced.[39] Regardless, based on simulations employing the Pitman hydrological model, there exists the "potential for dramatic changes to Okavango River discharge under future climate conditions, but with considerable uncertainty in the magnitude of any future changes."[40] Still other shifts are localized to particular ecological systems that experience seasonal, annual, and multi-decadal cycles of change. While the future is unknown and can only be projected by modeling or other forms of estimation, the likelihood is that global climate change will result in greater variability in biophysical patterns.

While there has been a renaissance of work on the interconnections between social and ecological systems, the specific relationships with livelihoods and human health are not always well understood. Over the past several years, I have been working with a team of researchers from the Pennsylvania State University, the University of Texas, and the Okavango Research Institute at the University of Botswana to examine ecological variabilities in the Okavango Delta associated with precipitation and flooding. The Okavango Delta, one of the most beautiful and biologically diverse environments in the world, exists largely because the Okavango River brings cycles of flooding to the delta. The Okavango River begins in the highlands of Angola, where it moves southeast to the border with Namibia and then flows into northern Botswana, into several main river channels that extend like an outstretched hand to the southern extent of the delta. Rainfall in Angola feeds the Okavango River as it flows into Botswana, where it is nourished by precipitation during the peak rainy season during the months of January and February. The floodwater arrives in the delta primarily in April and May, and while flooding events tend to be biannual, the exact timing, duration, amount, and location of these waters can vary considerably. For example, in certain parts of the delta, the extremely limited topography means that animal paths, such as those created by hippos, can alter the flow of the water from one year to the next. The extent of the delta can vary by as much as 50 percent within or between years, with some channels becoming dry for over twenty years and then filling and flooding. Precipitation also varies within the region by more than 100 percent from year to year, and over distances as short as ten kilometers.[41] The spatial and temporal variability of flooding and precipitation means that the setting is subject to dynamic change. This dynamism can be predictable and follow a pattern, but it can also be unpredictable and raise uncertainties for residents living in the area.

Over the past couple of decades, there has been a reduction in the amount of water in certain parts of the delta, which encouraged residents to settle in the secondary floodplain and establish homes as well as agricultural fields. Increases in precipitation in the past decade have increased the amount of water in the Okavango region, thereby challenging social systems in the region. The Okavango Research Institute reports that inflow at Mohembo, the inlet of the delta at the Namibian border, steadily increased from 9,800 mm^3/year in 2007 to over 13,300 mm^3/year in 2011.[42] Returning to my analogy of an outstretched hand, the center of the hand continues to feed the delta even when the floods recede in September and October, such that some areas have permanent water year round. The system is characterized by a variety of ecosystems, including permanent swamp, areas covered by sedge and grass vegetation (seasonal flood-plains), and floodplain grasslands (intermittently flooded areas).[43] The permanent swamp areas differ from the seasonal floodplains and inter-mittently flooded areas that experience the spatial and temporal variabil-ity in terms of the amount, location, and duration of water. When the channels are dry, settlement in the secondary floodplain has multiple ben-efits because it means closer proximity to water that is present in the sys-tem year round. In the Etsha region, where we have been interviewing local residents since 2010, there are tightly coupled relationships between social and ecological systems that can be reworked with variabilities in precipitation and flooding.[44] This can be seen through livelihood practices that depend on the water in various ways, including the collection of reeds and thatch grasses, fishing, farming in the secondary floodplain, and live-stock grazing.[45]

Increases in the amount of water resulted in human displacement, with some villages in the region being flooded completely. The government of Botswana reported that floods displaced more than 800 families in the Okavango Delta in 2009; another 170 families were relocated in 2010.[46] Fortunately for some of these residents, the Botswana government has been able to expend funds to relocate people by various forms of trans-port, including motorboats, and establish temporary settlements for them. Interviews with residents living in the village of Etsha 13 were conducted in 2011 and showed disproportionate impacts within the village itself.[47] Residents living in closer proximity to the secondary floodplain experi-enced flooding while other community members located farther away were not directly affected. Some of these interviews were conducted with community members who had been resettled in government tents to move them to other parts of the village.[48] In some cases, residents were resisting the relocation and were insistent on returning to their homes once the floodwater receded, which they anticipated would occur sometime in the next few years.[49] Regardless of the eventual outcome of these political environmental interactions, it is clear that the ecological variability associ-ated with precipitation and flooding has significant impacts for social systems in the delta.

One of the consequences of this dynamism is that rivers farther south received water after being dry for decades. One such river is the Boteti, which dried in the late 1980s as a result of shifts in precipitation and flooding patterns. This was viewed with some alarm, as evidenced by one governmental report that warned about desertification in the region.[50] Taking a longer view, the Boteti region is characterized by variability in precipitation and flooding, but the system has tended to follow a multi-decadal cycle. One study suggested that the region experiences a drought cycle with a periodicity of eighteen years, with a coinciding flow regime.[51] Villages farther downstream, including those toward the Mopipi Dam, which has been fed by the Boteti when it is fully flowing, stopped receiving water in the 1980s. Increases in the amount of floodwater in the Okavango Delta fed the Thamalakane River in Maun, which, coupled with increases in precipitation in 2008 and 2009, resulted in the river channel filling with water and slowly returning toward the Mopipi Dam in 2011. Figure 8 is a map of the Boteti study area.

The gradual return of the Boteti River presented an ideal opportunity to evaluate the coupled relationships between social and ecological systems and the ways in which these changes were affecting livelihood systems, human health, and well-being.[52] In May and June 2010, members of the research team conducted twenty interviews in the region. In the remainder of this chapter, I report on ten interviews that I conducted with individuals and families living directly alongside the Boteti. Working from the southernmost point where the water had reached and stopped flowing, interviewing was conducted at three distinct locations, first in areas where the water had yet to return, then where water had just returned that year, and finally in an area where water had been present for more than a year. Figure 9 shows the southernmost extent of the Boteti River.

The intention of the sampling design was to create a socio-ecological transect along the Boteti River, in which responses by local populations could be evaluated to understand their perceptions of ecological variability and to note the ways in which they were adjusting their livelihood systems either in anticipation, or as a direct result, of the newly arrived water. While I expected to learn much about how livelihood systems might be disrupted by shifting ecological conditions, I was also intent on under-

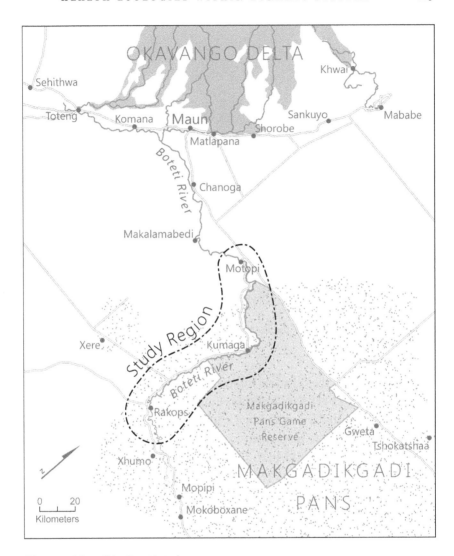

Figure 8. Map of the Boteti study area.

standing how the region's shifting political environmental context was influencing the domain of health for residents. It was likely that the possibilities for health and well-being were reflected in perceptions of environmental change that were tied to their age, residence time in the region, and assets that could be invested to adjust to environmental change. In

Figure 9. Water returning to the Boteti River (June 2010).

the Boteti region, these factors intersect in complex and dynamic ways to produce shifting health ecologies for human populations.

It was clear that residents had conflicting views about whether the water would return to their location alongside the river, in addition to the specific impacts it would have for them and others living in the area. Interviews with many residents revealed their awareness of the multi-decadal flooding pattern. It was common for older community members to talk about a natural cycle for the flooding occurring roughly every twenty years. Younger residents, especially those who grew up while the river was dry, had different understandings. Some respondents remarked that they didn't believe the water was going to come all the way to them, while others commented, with some disbelief, that the riverbed was filling again. Oral histories that gauged residence time and memories about the river showed some discrepancies, but in general 1987 and 1988 were the years mentioned by residents as those when the river was last seen at their locations. Respondents explained that the water returned in 2008 and then receded back into the channel, only to return and then stay in either

Figure 10. Cattle *kraal* in the Boteti region (June 2010).

2009 or 2010, depending on their location. In talking about the return of the water, one resident explained:

> I did not think the river was going to come. We heard rumors that the river was coming but I did not think it would come back. We made a borehole in the river because we did not think it was coming. It is like a dream because we thought the river was gone for good. The first time the river came it stopped at Makalamabedi.

The return of the Boteti River triggered livelihood adjustments for residents. For those able to invest in livestock production, which in the region is primarily cattle and goats, the availability of the water was indicated as a variable in expanding their herds and reducing costs for inputs. Flooding in the riverbed means that the boreholes, which are wells dug into the ground to access subsurface water, have to be removed. As we drove in an all-wheel-drive vehicle in the riverbed to the northern area where the water had stopped, many boreholes were operating with fuel-powered generators that would pump water for livestock and agriculture (figures 10 and 11).

Figure 11. Borehole in a cattle *kraal,* with the fuel-powered generator that brings water to the livestock (June 2010).

Surrounding the boreholes were multiple penned areas *(kraals)* where cattle were kept when not set loose for grazing in the open areas above the river channel. The return of the water meant, for some families, that their costs in terms of fuel would be reduced, thereby providing another incentive to invest in more livestock if they had the means. It should be noted that unlike the Okavango Delta itself, the Boteti River is a straight channel with enough topographic relief on either side that the water stays contained in the riverbed. This reduces the possibility of overflow from the river that would cause the kinds of displacement seen in the Etsha region. The geomorphology of the Boteti is particularly important in shaping the political environmental context because it generates a distinct socio-ecological dynamic. The consequence is that residents living along the river channel did not speak about the possibility of flooding on their compounds but instead were strategizing potential adjustments, such as the removal of a borehole or attempting to grow crops to take advantage of the newly arrived water.

The possibilities for human health are tied to agricultural practices that are shaped by ecological systems experiencing change. This means that the return of the Boteti River has direct impacts on food security, especially given anticipated future variability in precipitation and flooding due to global climate change. Some residents explained that they were attempting to take advantage of the water's return by securing seeds of various crops, including corn, pumpkins, cowpeas, watermelon, sorghum, calabash, spinach, and sugarcane. Others discussed how they were changing their cropping patterns to take advantage of the water's return. Some of the interviews were conducted with residents who engage in *molapo* farming that occurs in the secondary floodplain.[53] Molapo farming is a traditional form of agriculture that has been practiced for several hundred years.[54] In the Okavango Delta, farmers wait for the recession of the floodwater at the end of the wet season to plant seeds in the relatively moist and nutrient-rich soil. As a result, it is a form of agriculture that is highly dependent on the flooding regime to provide water and nutrients to the crops for production. Along the Boteti River, the location and timing of the molapo field is closely linked to its productivity because too much or not enough water can spell disaster for the farmer. One respondent named Fulane expressed her intention to plow by the river but said that she was going to wait to see what the water would do. Even if the water level continued to expand, she described her uncertainty as to whether it would cover the fields and be of any use; therefore, it seemed prudent to wait to see what the river would do over the course of the year. The return of the Boteti was not enough to trigger new agricultural practices for some residents; they explained that rain was also critical for farming, particularly because irrigation could be challenging. Fulane explained that the family planned to do one of two things: either plant a small field near the river with crops for household consumption or plant a small field with a larger one farther away that would grow crops primarily for sale. The investment of time and other resources into planting two fields was dependent on the flooding and precipitation patterns, and her family was waiting to see how these dynamics would unfold before deciding on their agricultural strategy.

Lastly, the collection and use of natural resources were also anticipated to shift in the future. In the villages in the Okavango Delta in which I have been working, collection and use of natural resources form a substantial part of livelihood production, with some households largely dependent on these resources for food, building materials, and income. Reeds and thatch are used widely in the region for the construction of homes and roofing, in addition to the production of baskets and other items for use or sale. Some respondents in the Boteti region indicated that they expected common reeds and thatching grass to become more abundant with the water's return. While interviews did not indicate that residents had begun to collect natural resources as of yet, likely because it was too soon, many spoke in anticipation of being able to do so in the future. Older residents reflected that these were practices they had done in the past and were looking forward to doing again when the resources became more widely available. Other residents indicated that they intended to fish or collect water lilies that are mixed with certain dishes for consumption. In these cases, ecological changes along the Boteti were directly affecting social systems, including livelihood production strategies such as natural resource collection, livestock grazing, and farming. The variabilities associated with precipitation and flooding, and the associated responses from human populations, reveal the tightly linked relationships between social and ecological systems that result in specific livelihood possibilities.

The social ecology of health approach frames human health not as the absence of disease but as the capacity to live well. Human health includes the attainment of material goods needed for quality of life and is a fluid domain produced through the dynamic interactions between social and ecological systems. The consequence is that the possibilities for health are interlinked with a gamut of social needs, including resources for livelihoods and foods that contribute to well-being. In terms of managed HIV, this means that HIV-positive individuals need access to nutritious foods to strengthen their immune systems and support the effectiveness of ARVs. In the Boteti region, food security is influenced by rising water levels that present opportunities and constraints for agricultural production. In some cases, families have material resources to invest in new types of agriculture and are thus equipped to benefit from shifting ecological patterns. In other cases, a lack of resources puts additional strain on the well-being of individuals and families in the region.

Uncertainty about the flooding and precipitation is a feature of life in the Boteti region. One resident discussed how the river was not like it had been in the past, suggesting that

> we believe in the ancestors. My grandfather told us that the river comes and goes away and can be dry. This happens every twelve years. But this river is not like when I was young. It is no longer flowing and we cannot depend on it. We used to fish and collect water lilies but right now there is nothing to do with that. It is not life water.

Some residents talked about the differences between the river as it existed in the present and the one in their memories. One man said the water was not enough for them and that they could not depend on it. In the past they could collect fish, water lilies, and reeds, but at that moment the river was not useful to him and members of his family. These interviews revealed fascinating connections between health and ecological dynamism in the region. The Boteti River exists materially but also symbolically in the ways that individuals understand their well-being. Water is the lifeblood for residents, and its ebbs and flows contribute directly to the domain of health. The variability in the water, whether seasonally or in decadal cycles, generates opportunities and constraints for social systems in the region.

I vividly remember many of these interviews, six years after visiting families living along the Boteti River. Two interviews stand out to me because they were conducted with members of the family of varying ages and revealed generational differences in terms of their understandings, and also memories, of the Boteti. Residents' perceptions of environmental change reveal much about the different ways they experience the water and the surrounding environment, as well as the changes they anticipate to their livelihoods and health in the future. Both of these families were highly dependent on livestock and agriculture and showed remarkable foresight in considering the ways in which they might be affected by the water's return. In the first interview, we visited with a woman about forty years of age who had moved to her home in 1982.[55] Esther remembered that just three years later the river had started to drop and had not returned to them, though they were expecting it to arrive soon. With some excitement in anticipation of the coming water, she explained that it would be a big change for their family. She expected they would start farming a variety of crops, including corn, sorghum, watermelon, beans, cowpeas, peanuts, and spinach. At this point we were taken on a tour of their cattle kraal and borehole, a short walk from their home. The family had roughly twenty-four goats, eight sheep, and perhaps a hundred cattle, according to the eldest daughter, Tsikani, though her mother laughed and said "maybe" to this estimate.

One immediate benefit of the river's return would be the water for their cattle, which was being pumped out of the ground using an electric motor. This is costly because the family had to travel to Rakops in their truck to collect fuel, sometimes going twice a month. In addition to saving on such costs, the water would likely taste better—there was a buffalo thorn root in the borehole, Esther explained, and the water pumped out had a bad flavor. The return of the water also meant the possibility that reeds would return and could be collected for use by the family. Lastly, Esther noted that thatch grass would become available, gesturing as she walked to where it would return farther up the bank. I was joined in the walk by Tsikani, who shared her experiences living next to the Boteti. Tsikani was insistent in pointing out that she had never seen the river in her lifetime and was uncertain that she ever would. She commented that "I have never seen it. I was born here in 1987, and I am not sure it will come. The older residents think it will come, and every year they say it will be next year. I don't think it will come." Tsikani laughed when detailing how residents had differing opinions about the return of the water and what it would mean for them. When I asked her mother's opinion, she said, "I don't know. We wait for it every year in June and it does not come."

Given the importance of livestock to the family, the subject of lumpy skin disease came up quickly in our conversation. It was shared that African cape buffalo drank from the water upstream, and this contributed to the spread of the flies that bring the disease. Time and again, residents raised concerns about lumpy skin disease and what it would do to their cattle. The disease can cause chronic debilitation similar to that experienced with foot-and-mouth disease. Severe and permanent damage to the hides of cattle can result, as can lesions in the mouth, pharynx, and respiratory tract. Because of these symptoms, including the possibility of mortality, lumpy skin disease can have serious economic consequences for families.[56] In spite of this threat, the family was awaiting the return of the water with excitement, anticipating that their costs of living would be reduced by greater opportunities to access resources in the future. As the interview concluded, they all piled in their truck to drive to Rakops to complete some errands, having already lost valuable time by talking with us for nearly two hours.

The second memorable interview occurred farther north, where the water had just arrived the previous year. We spoke with Joshua, a man about sixty-five years old, outside his small home near the Boteti River. Also present was a young child, my colleague Dr. Kenneth Young, and our interpreter Kgosi, who was there to assist with the translation from Setswana to English. Joshua explained that he had worked in a mine at a younger age and had left without receiving a pension or adequate medical care from the hospital in the town of Maun. He was born at that plot in 1953, and he remembered flowing water with fish, water lilies, reeds, and papyrus. But the water had eventually left, and the riverbed was dry for fifteen years before it started flowing again. When asked if he had thought the river would return, Joshua said no—he didn't believe it would. But the water had arrived in 2008 and receded; had returned in 2009 and stayed; and it remained there when we arrived to visit with him. When he first heard from others that the water had returned farther north, he rode his donkey along the riverbed to see the water. When he returned to the house, he told his sons to remove the bore-hole, but the water didn't arrive that year. Joshua thought then that it wouldn't come to his stand, so he rode again up the river to Xaega, where the water was settled in lagoons. Livestock were a major feature of his household economy: more than eighteen cattle and fifteen goats were owned by the family. In 2006, Joshua took his cattle to Xaega, where his sons were living on a cattle post. He noted that he didn't plan on bringing the cattle back, because he thought they could get lost or eaten by wild animals such as hyenas. The family had a field of seven hectares that was being used to grow sorghum, sugarcane, pumpkins, beans, and corn. He explained that his farming was dependent on both flooding and rainfall, with one section of the field located adjacent to the riverbed and the remainder farther away and thus reliant on precipitation. Different crops are grown depending on the flooding and the rainfall, thereby linking these two processes to ensure the production of food for his family.

In comparing the current river with the one he remembered from the 1970s, Joshua described the earlier river in more nostalgic and romantic ways. He explained that the current river was not as good because it was not producing the resources they could use, such as papyrus for the lower layer of his thatch roof to keep the house cooler in summer. Joshua also noted there were no water lilies or reeds at that time, emphasizing that nothing was "germinating" in the riverbed. When asked about the gradual return of the Boteti, he said that he didn't think people should use the resources until later. Joshua explained that he would not fish in the lagoons because the fish were small and they should be allowed to grow bigger and regrow with the flowing water. He also thought people should not pump water out of the river to irrigate their crops, because it leaves the water dirty for others. I was curious what he thought would happen in the future and, like many others with whom I spoke, he insisted that the future is unknown. Joshua explained that the river might stay but that it also might not, remarking that this is part of the cycle of life along the Boteti. In my field notes, I reflected that Joshua seemed very pragmatic compared to some of the other residents with whom we had spoken, carefully considering the slow return of the water and the potential consequences for himself and his family. This underscores that the individual circumstances for each family play an active role in shaping the ways in which the ecological shifts in the region, particularly precipitation and flooding, affect livelihood systems, food security, and health and well-being. It is these dynamics— whether they are based on age, gender, residence history, agricultural knowledge and flexibility, access to resources, or economic inputs—that contribute to shaping how health is embodied, experienced, and managed along the Boteti River.

HEALTHY RIVER, HEALTHY LIVES

The return of the Boteti River shapes livelihood systems and perceptions of health and well-being in complicated and diverse ways. Some respondents discussed the health effects of the water's return, expressing concern about diseases that would affect people and livestock. As already noted, a common complaint was lumpy skin disease, which infects cattle in the area. Residents differed on the source of the disease and whether the return of the water was intensifying the rates of incidence. In two interviews, older residents who had lived in the area longer disputed the contention that lumpy skin disease came with the river. In one case, an elderly gentleman insisted that he had never heard in his childhood that diseases come with the river, and that lumpy skin was already present in the 1970s. Others spoke about how their grandparents had lived with the river and there were not diseases at that time. Still another woman said that it was the rains that brought disease because they caused the grasses to grow taller, which is what causes flies to appear. Other interviews suggested that mosquitoes would become more prevalent along the Boteti, and this would mean diseases for people as well as livestock. Beyond the significance of these contrasting views is the fact that within these communities there are multiple interpretations of disease vectors and health vulnerabilities to people and cattle. In addition to waterborne disease, informants discussed how children would be at risk because crocodiles might return with the water. One respondent predicted that children would die trying to swim in the river because of encounters with crocodiles and hippos.

Yet health is not understood strictly in terms of exposure to infectious diseases such as lumpy skin or malaria. Residents routinely talked about the return of the Boteti River as having symbolic and cultural importance for health and well-being. They referred to the water as a source of life for families, their cattle, and the natural environment more broadly. Drawing upon my earlier discussion, Boteti lifeways are intimately connected to the functioning and health of the natural environment. One elderly man, who had lived in the area since 1955, explained that the return of the water "will be good because water is life. When there is no water then there is nothing." Regardless of these positive sentiments, however, he spoke of how hippos and crocodiles will come with the water and how mosquitoes will arrive to lay eggs and spread malaria. Another man said that it would be good for their family if the river went past them downstream, explaining that if it stopped immediately next to their home, the water would not stay clean and there would be mosquitoes and infection for the cattle. Still another resident explained that the return of the water would be positive for precipitation because the water will evaporate and improve the rains for the agricultural fields. Regardless of these views, multiple interviews revealed that the flow of the water was understood as a "natural" event. As one person said, "The river will come and the river will go. No one knows what it will do." This coupling of the anticipated positive impacts with negative concerns was prevalent in many of the interviews and suggests a multifaceted perspective on the health domain within the shifting political environmental context of the Boteti region. Dynamic biophysical conditions mean new livelihood possibilities for some that have the potential to improve health and quality of life. The uncertainties associated with environmental change also mean that other families will face new challenges in meeting basic needs.

Fluctuations in the ecological conditions along the Boteti River produce shifting health ecologies over time and space. Following the riverbed to the north helps reveal how the political environmental context plays a determinative role in producing human health and vulnerabilities to disease. As evidenced by these interviews, the return of the water is not simply a biophysical process. The Boteti River filters through political, economic, generational, gendered, and cultural systems that enable and constrain the ways in which individuals and families respond to shifting conditions. This is reshaping social systems and livelihoods in various ways, ranging from the dismantlement of boreholes to new agricultural practices to anticipated increases in natural resource collection in the future. This is pressuring existing social systems while presenting opportunities for residents who have the means and capacities to adjust their livelihood systems. These shifting ecologies are multifaceted and complex; they can be positive for some and negative for others. For those fortunate enough to possess capital to offset the disruptions from moving and replacing boreholes, the Boteti River means a reduction in expenses that can be directed to other needs. While residents spoke openly about the return of the water as a "natural thing," the ways in which it subsequently shapes human health should not be simply reduced to a natural process.[57] In this case, changing biophysical conditions have a direct bearing on the domain of human health.

Health and well-being have the potential to improve should the increased flow of water contribute productively to agricultural and livestock production. More livestock could mean more resources available to offset school fees, medical costs, or the purchase of other items that improve food security. Increased and diversified agricultural production could mean more nutritional content and caloric inputs to individuals that might offset the risks of infectious diseases and improve quality of life for family members. The expansion of the human health domain is not uniform along the Boteti. Residents with the means to make adjustments are in a better position to take advantage of these shifting conditions. Much like their neighbors upstream or down, they navigate a political environmental context that is variable—and hence their individual and family health prospects are similarly variable. These health ecologies will experience change as the social and ecological systems of the Boteti region, and those of the Okavango Delta more broadly, continue to shift. The life-ways and landscapes of the Boteti region provide a window into broader processes of ecological dynamism that will have significant impacts for environmental health and human well-being in the future.

6 States of Health

It was a cloudy and breezy day when I first met with the Mzinti home-based care group in their small office in November 2012. Because this was the beginning of the research project on HIV–environment interactions, it was important to introduce my research team, obtain permissions from political authorities, and request input about information that would help guide health-care efforts in South Africa. This was our first interaction with this remarkable group, but one that would be repeated in future years. Like its neighboring village of Ntunda, Mzinti has one home-based care organization that works within the community, providing counseling and referrals for testing, offering routine services such as cleaning, and supplying hospice care for end-of-life patients. In my time in the community, I would see these volunteers walking to their assigned houses with clipboards and backpacks, checking on children and on those managing illness. When we arrived at the office that day, several dozen people were waiting outside of the small building that is located near the center of the town. Immediately apparent were the different colors of the uniforms being worn, which distinguished two community organizations. The first was an agricultural group that provides extension training to support local farmers. It is likely that their job has become more challenging in recent years because of the expansion of commercial development in the region, including in areas adjacent to Mzinti. Although the community has historically been surrounded by communal space where natural-resource collection and livestock grazing are allowed, these areas are under pressure for other economic initiatives. In addition to territory that was committed to sugarcane development in 2001, land directly north of the main tarred road is being converted to a shopping mall, and these types of initiatives have continued apace in recent years. The consequence is that the surrounding communal areas in which residents are able to invest in agricultural development are shrinking as the population expands. There were about fifteen members of the agricultural extension group, and all were wearing orange uniforms with matching pants and shirts. By comparison, many members of the home-based care group were dressed in blue uniforms. In addition to the divergence in uniform color, there was a clear gender division. The agricultural organization was nearly all male, and the home-based care group consisted entirely of women.

I returned to the Mzinti home-based care group with members of my research team in June 2014 to hear more about the challenges they face in providing health-care services. The director of the organization, Buyisiwe, stressed how difficult it was to meet needs in the community with the limited resources they receive from the government. She shared that they had searched for outside monies, including applying for funding from the U.S. Agency for International Development (USAID) Sexual HIV Prevention Program. They were not successful; however, they had learned that the home-based care group in Ntunda had received a grant from the agency. The Mzinti group had been receiving support from the national government, but this had ended several months earlier and they were actively trying to find new sources of funding. Our visit occurred just two months after the national elections, which were historic because they took place twenty years after the 1994 democratic elections. I could not help but notice Buyisiwe's attire: a bright yellow ANC shirt with a picture of Nelson Mandela on the front, accompanied by an ANC head-scarf. In the weeks I had spent in the region, the black, yellow, and green of the ANC were visible throughout the villages, largely in the clothing worn by residents. Many were adorned with a picture of a smiling President Jacob Zuma.

The results of the national elections were still being discussed in the country because they signaled potentially momentous changes in the political landscape. Although the ANC retained its control over Parliament, there were signs of slippage in the form of reduced power in the recent elections and greater talk about opposition parties and the need for new leadership. Notable in the elections were the "Born Free" generation, which were newly eligible voters who were born after the 1994 democratic elections. For the first time, voters participated who had no personal experience with apartheid. Support for the ANC remained strong in the country, particularly in some of the rural areas, though there had been inroads in Gauteng Province by the Democratic Alliance. A second opposition party, the Economic Freedom Fighters (EFF), had been successful in aligning with the interests of the poor and working classes, which are groups that have typically supported the ANC. In a media-relations coup, nine EFF members were sworn into Parliament in May wearing bright red service-worker uniforms. *Corruption* has become a keyword for criticizing the ANC. Younger residents, including members of my research team who have lived in these villages their entire lives, indicate less commitment to the governing party and express concerns about future opportunities. Before we entered the Mzinti home-based care building, Elinah shared that she didn't vote in the elections because she didn't see the point, explaining that young people are frustrated—and those who did participate voted for the ANC, and not for President Zuma. Elinah emphasized that there were not long lines on election day as in the past and that people were finished by early evening. Upon hearing this, Soninwe became animated and responded that things do not change unless you participate. Later, when returning to Ntunda after a day of work, Soninwe shared with me her frustration about how the government had failed to provide services and unfairly administers development programs. The twenty years that had passed since the democratic elections served as a milestone in many ways, including coinciding with a new generation of voters with different concerns and commitments.[1]

Although funding support to the Mzinti home-based care group had stopped in May, Buyisiwe spoke of new ideas for providing services to residents. She indicated an interest in building a kitchen to provide "good healthy food" for patients waiting at the Mzinti clinic to receive care. The clinic is on the other side of town, and she emphasized that it would be convenient for patients to come and take advantage of this service, particularly patients who have to wait all day and then return to their homes. Buyisiwe explained that this would be especially helpful for those living in the RDP homes, the government-built subsidized housing development on the eastern edge of the community. Buyisiwe proudly noted the completion of the fence that had allowed them to open a garden to grow vegetables to share with families that they visit. The importance of a healthy lifestyle was emphasized as we took a tour to see the different crops being grown for residents in the community.

These encounters demonstrate that human health is produced and experienced through interactions between social and ecological systems that are generated by historical and spatial processes. Only by addressing the coupled and dynamic intersections between these systems is it possible to understand the production of health and differential vulnerabilities to disease. Because the political environmental context changes over time, it creates inequitable outcomes in terms of which populations are exposed to health risks and the conditions that undermine well-being. The political environmental context results in dynamic systems that transform lifeways, establish new patterns of disease, and produce disproportionate exposure by race, ethnicity, gender, socioeconomic class, and location. The political environmental context creates the states of disease for individuals, families, and communities. Spatial processes shape the ways that health is embodied, experienced, and managed.

My discussion has been centered on the social ecology of health framework, which offers three critical contributions. First, the framework attends to how historical spatial formations contribute to shaping contemporary vulnerabilities to disease or exposure to social and ecological factors that produce poor health. The expansion of the HIV/AIDS epidemic in South Africa occurred in a particular context that facilitated its transmission in distinct ways. The continued uneven rates of HIV infection are testament to this and demonstrate that the landscapes of HIV are not two-dimensional; rather, they have been differentially produced through mechanisms of spatial engineering that continue to be expressed in terms of access to information, social services, testing, stigma, and the provision of ARVs. Interviews conducted with members of home-based care groups reveal much about the ways in which they provide services while navigating local dynamics and national bureaucracies. The needs of patients extend beyond the management of a particular disease to include resources necessary for maintaining health and well-being. For example, some volunteers from these groups shared that in their regular encounters with patients, it is common to find homes without food. Patients complain to them about lacking resources to meet basic needs, which is compounded by the challenges raised by managed HIV. Health care is a multifaceted interaction that includes attention to food security, particularly for patients who are taking ARVs and thus have additional caloric and nutritional needs. Visits to clinics for blood testing to measure CD4 counts and collect ARVs are accompanied by messaging about the importance of minimizing stress while maintaining a healthy diet. When resources are available, home-based care volunteers make porridge to ensure that patients have enough food to handle treatment. The lack of food is a major reason for failing to adhere to ARVs, which was echoed by one home-based care volunteer, who explained:

> My problem is that when we go outside to visit patients we find a lot of problems, like patients are taking treatment so they need food but they are unemployed. Social workers promised to give them food so they must be registered. After registration they disappear. No food. We are also suffering. We are volunteers and have no money. Please help us. We have a garden. We give vegetables to our patients . . . because they take pills and they deserve those. We will give them spinach to eat. Some do not have houses to sleep in. Even those of us at the home-based care have problems. Please help us to develop our center. If we can get one hundred dollars or even fifty dollars that is enough for us so that we can be able to buy bread because the patients see us as liars. So now we are afraid to promise them anything because of the past experiences with social workers and other organizations. They will think we are the same as the others.

This statement speaks to the political and bureaucratic challenges of engaging with and caring for sick individuals in the era of managed HIV. Resources are scarce and intentions are misunderstood, which means that the full gamut of needs are not easily addressed. Biomedical HIV is about testing and accessing ARVs, but HIV lifeways involve much more. Managed HIV is not simply about accessing life-saving drugs—it is also about securing other resources needed for survival. As has been noted by other researchers, the presence of HIV in the body has taken on a new course, with hunger becoming the embodied experience and central to the daily needs of infected individuals and those attempting to provide care. In ethnographic research from Mozambique, Ippolytos Kalofonos detailed the mechanisms through which the infected access external sources of support, navigating bureaucratic systems in their efforts to secure food.[2] Kalofonos showed how social practices are affected, while simultaneously being mediated, by national and global political and economic systems.[3] The home-based care groups that provide daily services to community members must similarly navigate political and spatial economies that operate at various levels and influence the information they have about the population, the availability of basic resources for caregiving, and political environmental constraints on livelihood possibilities and food security needed to support managed HIV.

Managed HIV is not solely about accessing desperately needed drugs; it is an embodied experience that is shaped by historical and spatial processes that determine which people have access to resources needed for survival. The use of these drugs occurs on a daily basis within multilayered and multiscalar political economies that are developmental, bureaucratic, and pharmacological. Community members navigate an array of health-care centers, whether they are clinics, home-based care groups, private facilities, or hospitals. These encounters generate particular interactions with the state and its discourses on biomedical HIV that do not neatly fit the multifaceted understandings that exist in these communities. Whether at the traditional medicine market or on the street corner where billboards are displayed, local understandings of disease can conflict with national representations. They remain place-specific and are rooted in historical dynamics that have been engineered by colonial and apartheid governments. Spatial processes have produced the contemporary landscapes of HIV that generate differential vulnerabilities to disease and often rigid lifeways through which people attempt to respond. Differential vulnerabilities to HIV infection, coupled with political and economic challenges in accessing ARVs, demonstrate how the political environmental context shapes human health.

Yet these relationships are not simply political or economic; they are also ecological. Access to natural resources, land allocation and tenure systems, and caloric and nutritional demands all intersect with biophysical processes that shape the trajectories of health and well-being. Expanded HIV access has become a driver for food security, such that the systems of land tenure that determine who is able to produce food become central to HIV lifeways. Land ownership in South Africa is gendered because women have not had equal rights to property. In the event of the death of a male head of household, previous practices meant the absorption of the family and land within his extended family, thereby reducing the ability of women to control their own economic destinies. Livelihood practices in this context are also gendered, with women largely maintaining households and children while overseeing household garden plots and cooking. Men are primarily responsible for livestock and agricultural production, when it exists, outside of the household property. The collection and use of natural resources are similarly bifurcated, with women often collecting water while men collect wood and other natural resources. This, in turn, means that encounters with the natural environment are gendered and are fundamental in shaping health–environment interactions in the region.

Evaluating the production of human health across multiple settings makes it possible to see how these patterns are social, ecological, and spatial. In the previous chapters, I referred to the spatial economies that have produced distinct vulnerabilities to HIV. Similarly, the spatial economies that produce landscapes of differential exposure to toxins and pollutants in the United States generate distinct health outcomes for communities. The anticipated effects of global climate change will be significant for environmental health and for the interactions between human populations and the natural environment. Changing weather patterns and environmental variability associated with climate change are anticipated to present constraints as well as opportunities for populations in ways that influence the boundaries of the health domain for individuals, families, and communities. Shifting ecological conditions associated with climate change and other weather-related dynamics are transforming the interactions between human and nonhuman worlds in ways that will have lasting implications for the spread of infectious disease and the maintenance of health. The return of the Boteti River serves as a microcosm of the interconnected relationships between health–environment interactions. While the water's return presents a livelihood opportunity for many in the region, concerns about lumpy skin disease and malaria shape many understandings of health and well-being. The health ecologies of the Boteti are shifting in ways that show the fluidity of human health. Environmental change in this setting is dynamic, yet Boteti lifeways are created by spatial systems that are rooted in political, economic, cultural, and ecological patterns.

The interactions between social and ecological systems serve as a central point of analysis for multiple research fields, including coupled human and natural systems; sustainability science; socio-ecological systems; and live-lihood studies.[4] This work is based, in part, on a conceptualization of dis-ease as a type of shock that has large, unpredictable, and irregular impacts.[5] But various diseases and health vulnerabilities occur over extended periods and have variable outcomes for human populations and ecosystems. Ecological changes can trigger adjustments and increase vulnerabilities, depending on the stability of the systems in question. Additionally, the role of political and economic systems in producing human health is not always addressed by the scholarly literature, and hence this work can underemphasize these underlying structures. The social ecology of health framework emphasizes social and ecological determinants and their interactions in producing health outcomes within various settings. In so doing, it provides an approach that more fully integrates social and ecological perspectives in order to understand the diverse factors shaping the spread of infectious disease and differential exposure to noninfectious disease.

Moreover, the framework considers health in a holistic manner. Drawing upon the WHO's definition, human health is a right that is achieved when there is physical, mental, and social well-being.[6] Health is an expansive ideal that considers not only the existence of disease in the body but also the achievement of well-being and the capacity to live well. Health is achieved when people have the ability to generate a healthy life and is differentially experienced when there are inequities in vulnerabilities to disease or exposure to conditions that produce poor health. I began the book with the provocative statement that no one is ever healthy. Rather, we exist in a state between health and disease. Simply put, the embodiment of a particular disease does not encapsulate the totality of our existence. The emergence of managed HIV in South Africa speaks to this, in that a previously devastating disease has become more normalized and part of the daily lifeways for many in the country. To identify HIV-positive individuals as either "sick" or "well" does not represent the full extent of their situation, or their experiences navigating the landscapes of HIV. The result is that competing understandings of health reveal important details of the micro-politics and inequities in power that shape access to information, resources, and opportunities that influence health management. This means that human health is tangible but also situated, relational, contingent, and dynamic. Health is similar to the concept of justice, in that it is an ideal rather than a state of being at a particular moment in time. Human health is about possibilities. It is having the capacity to make decisions that result in a better and more secure quality of life.

The WHO's inclusion of mental and social well-being resonates with my discussion of managed HIV in South Africa. The mention of mental health is important and can be overlooked when facing threats from infectious disease and epidemics. Yet, when I spoke with residents in the Mzinti community in May 2014, many HIV-positive people talked about the strain they were under in managing their health and ensuring that younger family members have even their basic needs met. I remember speaking with Kagiso, a female resident of the RDP housing project, about her daily challenges. While HIV was very much part of her concerns, Kagiso instead emphasized the stress that she faced in providing for her children. In her case, HIV had receded into the background and, while it was still present in her sense of health, other concerns were on the surface. In a second interview with Nomusa, an HIV-positive woman in the same neighborhood, she broke down in tears when describing the ways she felt that others in the community were judging her. An umthandazi (a religious faith healer), Nomusa explained that the visits she received from some community residents made her neighbors feel that she was wealthy by comparison. Yet she insisted that she did not charge her patients and that they would only occasionally leave her a nominal gift as a token of gratitude. For these two women, their health and well-being cannot be distilled into the presence or absence of a particular virus. Rather, their sense of health is based on their unique circumstances, including their location within social networks and interpersonal dynamics that contribute to their identity and livelihood possibilities. These competing health narratives challenge conventional disease orthodoxies and representations produced by the biomedical model and by powerful institutions, including national and international agencies. Examples of these patterns have been described in other settings and underscore the urgency to unpack the multiple meanings and practices that shape health and well-being.[7]

Finally, the social ecology of health framework demonstrates the importance of spatial processes in producing disease vulnerabilities and the opportunities for health management. While scholarship in the social sciences has been insistent on the underlying structural conditions that produce human health, the role of spatial processes needs to be included in these analyses. The emergence of the U.S. environmental justice movement was based on the certainty that the production of space results in differential vulnerabilities to the spread of infectious disease and exposure to noninfectious disease. Additionally, the capacity to live well through access to green space for contemplation and recreation, or to maintain food security, remains unevenly experienced in urban and rural landscapes. While these cases show that human health is contextual and relational, they help demonstrate how the political environmental context shapes human health and contributes to producing vulnerabilities to disease and the conditions that produce poor health.

Because the political environmental context changes over time and space, it can be highly unequal in terms of who is exposed to health threats and the conditions that undermine human health. Health landscapes, whether in South Africa or northern Botswana, are deeply rooted and have been generated by political, economic, cultural, and ecological forces that result in inequitable exposure to disease conditions and access to the services needed to maintain well-being. Human health is produced through spatial conditions that are created over time to the benefit of certain actors. Sometimes these actions are overt and intentional. At other times they are subtle and hidden. But they have the same outcome in generating places and landscapes that provide differential vulnerabilities to the conditions that produce disease or opportunities for well-being. It remains necessary to engage with these dynamics to understand the multiple factors that intersect in producing the states of disease. Only by doing so is it possible to create more equitable and lasting states of health.

Notes

PREFACE

1. Belluck, 2009.
2. Dugger, 2009a.
3. World Health Organization, 1946 (cited in Grad, 2002, p. 984).
4. Farmer, 1999.
5. Farmer et al., 2013.
6. Economist, 2011.

INTRODUCTION

1. Quammen, 2012.
2. Dionne, 2014.
3. BBC News, 2014.
4. United Nations News Centre, 2015.
5. See de Waal and Whiteside, 2003; Drimie, 2003; Love, 2004; Murphy et al., 2005; Negin, 2005; Barnett and Whiteside, 2006.
6. Hunter et al., 2007.
7. See Platzky and Walker, 1985. The apartheid government deliberately undercounted rural populations in an attempt to minimize the perceived effects of its policies, in order to sway global public opinion. As a result, historical population censuses are notoriously unreliable.

8. Readers familiar with European colonial history will recognize Rudyard Kipling's famous poem satirizing American imperialism in the Philippine Islands. Titled "The White Man's Burden," it begins:

Take up the White Man's burden—
Send forth the best ye breed—
Go, bind your sons to exile,
To serve your captives' need.

There are echoes of these sentiments in colonial and apartheid interventions in South Africa: the state justified racial segregation as being in the best interests of the entire country, including the majority African population. Apartheid was founded on the principle that racial groupings could not coexist peacefully and, thus, spatial segregation was presented as necessary to ensure social harmony. In reality it was a mechanism for accumulation by dispossession (King, 2004, 2007).

9. Malan and Hattingh, 1976, p. 6.

10. South Africa Information Service, 1973, p. 30.

11. Fifty semistructured interviews were conducted between October 2001 and August 2002. The majority of these were completed in SiSwati, and I had invaluable assistance from Erens Ngubane, my research assistant, with the translation and transcription of these interviews. Other interviews were conducted in English, primarily with younger community members who had received language training in school, and in Portuguese with residents who had left Mozambique, in some cases fleeing the civil war that lasted from 1977 until 1992.

12. I have argued elsewhere that livestock owners in the community have been able to invoke the "cultural" importance of livestock to the Swazi population to make claims over territory and resources in the contemporary period (King, 2011). In a specific example of this, livestock owners in Mzinti were able to convince the tribal authority to demarcate land in the communal area for grazing space and an abattoir. This was in spite of the fact that a minority within the community, roughly 10 percent, owned cattle.

13. A return visit in 2012 revealed that Mahushe Shongwe Reserve managers were again allowing community residents to remove a circumscribed amount of wood for funerals.

14. See Makgoba, 2002; Nattrass, 2004; Fassin, 2007.

15. See Jones, 2005. The High Court of South Africa ruled in 2002 that the government needed to make nevirapine available to pregnant mothers to reduce transmission of HIV to their children. The Health Ministry took until 2003 to approve a provision plan, and the plan did not take effect in Gauteng Province until March 2004.

16. Ibid., p. 426.

17. See van der Vliet, 2004; Jones, 2005.

18. The contrast with the country of Uganda is noteworthy. By this time Uganda had launched its ABC program—which emphasized abstinence, being faithful, and using condoms—to worldwide acclaim and success in reducing HIV prevalence. For a discussion of the contrasts between these two countries, Thornton (2008) is helpful, particularly in arguing for the role of a strong governmental response in managing the epidemic. He credited, in particular, President Yoweri Museveni for "integrating the struggle against AIDS into local cultural forms and practices" (p. 140). The government's "Zero Grazing" and "Love Carefully" campaigns targeted multiple sexual partnerships and spatially expansive interactions.

19. Wines, 2006.

20. Campbell, 1997, 2003.

21. See Marks, 2002; Fassin and Schneider, 2003.

22. I have spent several years trying to find concrete evidence that would support these purported estimates, but to no avail. As such, they should be seen as largely conjectural but important in revealing discursive understandings of the extent of the epidemic. I comment in chapter 2 that these types of estimates continue to circulate as statements meant to challenge national estimates, in essence serving as a mechanism for criticizing the national government. Their circulation says much, not only about the potential severity of the epidemic, but also about the perspectives of social actors in the region.

23. Reported in Hunter, 2010b.

24. Acknowledging President Zuma's approach is not without complications. Before his presidency, while he was an ex–deputy president, he was put on trial in 2006 for the rape of a friend's daughter who was known to be HIV-positive. During cross-examination, he responded that even though he had not used a condom, he was unafraid of becoming infected because he had taken a shower afterward. His statement was immediately denounced by AIDS activists and health experts.

25. Chigwedere et al., 2008.

26. Ibid., p. 410.

27. Boko et al., 2007.

28. Paaijmans et al., 2010.

29. Lambrechts et al., 2011.

30. World Health Organization, 1946 (cited in Grad, 2002, p. 984).

31. Grad, 2002, p. 984.

32. See Lalonde, 1974. The Honourable Marc Lalonde was later recognized by the Pan American Health Organization as one of the leading figures in the field of public health in the twentieth century.

33. World Health Organization, 1986.

34. Butler, 2005.

35. Parkhurst (2004) is helpful in addressing the political environment of HIV in Uganda and South Africa, though the approach is markedly different from the social ecology of health framework that I outline in chapter 1.

36. Smith et al., 2014.

CHAPTER 1: SOCIAL ECOLOGY OF HEALTH

1. See Associated Press, 2008a. Cholera is an acute, diarrheal illness caused by infection of the intestine with the bacterium *Vibrio cholerae*. According to the Centers for Disease Control and Prevention, an estimated three to five million cases and more than a hundred thousand deaths occur each year around the world. Approximately one in twenty (5 percent) of those infected will have severe disease, characterized by profuse, watery diarrhea; vomiting; and leg cramps. This can lead to loss of bodily fluids, dehydration, and shock; if untreated, the severe form can cause death within hours (www.cdc.gov/cholera/general/). The bacterium is typically transmitted through water contaminated by fecal matter from an infected individual.

2. Bearak, 2008.

3. CNN, 2008.

4. BBC News, 2008.

5. Smith, 2009.

6. Associated Press, 2008b.

7. See Scheper-Hughes, 1992; Farmer, 1999, 2005.

8. This is akin to Amartya Sen's now famous contention that famines do not occur within functioning democratic systems, because of the existence of mechanisms that force responses by state agencies (Sen, 1999).

9. In describing the spread of disease within postcolonial Tanzania, Meredeth Turshen (1984) noted: "Disease agents such as bacteria and viruses may occur spontaneously in nature, but there is nothing natural or spontaneous about epidemics" (p. 15). Turshen's analysis concentrated on capitalist forms of production, made possible under colonialism, that reworked social systems such that vulnerabilities to poor health were increased. Her work examined the gendered disparities that resulted from agrarian policies, to show how their production was deeply rooted within past governance patterns.

10. Both Carter (2012) and Nash (2006) have provided compelling accounts of how the supplanting of miasmatic theory by germ theory in the twentieth century was integral to the management of malaria in Argentina and the United States, respectively.

11. Webb and Bain, 2011.

12. Ibid., p. 13.

13. In a detailed history of this case, McLeod (2000) noted that the cholera outbreak was already in decline by the time the Broad Street pump was removed; regardless of its accuracy, she noted, the Snow legend has become ubiquitous and is propagated by numerous academic disciplines.

14. Mayer and Pizer, 2008, p. 15.

15. See Mansfield, 2008; Crews and King, 2013.

16. See Mishler et al., 1981; Krieger, 1994.

17. Wade and Halligan, 2004, p. 1398.

18. Turshen, 1977.

19. Turshen, 1984, p. 11.

20. Farmer, 1999, p. 10.

21. I should note the important distinctions between conceptualizations of "disease" and of "illness." It is common in the literature to refer to illness as the subjective experience of symptoms, whereas the medical practitioner reviews those symptoms to "identify the specific underlying pathology in the patient's body that is producing the signs and symptoms, distinguish it reliably from other possible diagnoses, and label it correctly with the name of a medically recognized *disease*" (Davey and Seale, 1996, p. 9 [their emphasis]; cited in Gatrell and Elliott, 2009). Disease, then, is the medically recognized malady, and illness is how it is experienced. In some cases a particular illness is not recognized as a disease by the medical establishment, either because it has an unknown etiology or as a result of cultural circumstances.

22. Baer and Singer, 2009.

23. See Singer, 1996; Baer and Singer, 2009.

24. Stillwaggon, 2006.

25. Ferguson, 1990.

26. Mitchell, 2002.

27. Goldman, 2005.

28. Li (2007) asserted that forms of trusteeship in Indonesia include intentions that are "benevolent, even utopian. They desire to make the world better than it is. Their methods are subtle. If they resort to violence, it is in the name of a higher good—the population at large, the survival of species, the stimulation of growth. . . . They blend seamlessly into common sense" (p. 5).

29. Rostow, 1960.

30. See Wainwright, 2008. Citing Spivak's articulation of a concept that "we cannot not desire," Wainwright (p. 10) argued that "we cannot *not* desire development" (his emphasis). If health is understood as an ideal rather than a state of being, it has a similar resonance. For how is it possible that we cannot *not* desire health?

31. I addressed this in greater detail in King, 2013. Some examples in the social sciences literature include Davies, 1996; Ellis, 2000; Dercon et al., 2005; and Kgathi et al., 2007.

32. Ellis, 2000.

33. Ibid., p. 40.

34. The United Nations (2008) recognizes a "slum household" as a group of individuals, living under the same roof, who lack one or more of the following conditions: access to improved water, access to improved sanitation, sufficient area for living, durable housing, and security of tenure.

35. United Nations, 2013.

36. Turshen, 1984, p. 4.

37. Farmer et al., 2013.

38. Ibid., p. 320.

39. Murray and Lopez, 1997 (cited in Farmer et al., 2013, p. 323).

40. N'Goran et al., 1997.

41. See Faber, 1993; Harrison, 2011.

42. Lerner, 2010, p. 3.

43. Adams, 2001, p. 307.

44. Farmer et al., 2013, p. 316.

45. Beck, 1979.

46. Ibid.

47. Ibid.

48. New York State Department of Health, 2009.

49. Prohaska, 2014.

50. Bullard, 2007.

51. Environmental Protection Agency, 1994.

52. Bullard, 2005.

53. Ibid.

54. Ibid.

55. Bullard, 2005, p. 60.

56. Bullard, 2002, p. 35.

57. Hillier, 2005.

58. Ibid.

59. Walker et al., 2010.

60. Cotterill and Franklin, 1995.

61. See Chung and Myers, 1999; Kaufman, 1999.

62. Neighmond, 2014.

63. See Heynen et al., 2006; Dai, 2011.

64. See Centers for Disease Control and Prevention, 2005. National Health and Nutrition Examination Surveys showed that, by race and ethnicity, non-Hispanic African Americans and Mexican Americans had higher percentages of elevated blood lead levels (1.4 percent and 1.5 percent, respectively) than non-Hispanic whites (0.5 percent). Among subpopulations, non-Hispanic African Americans aged one to five years and above sixty years had the highest prevalence of high blood lead levels.

65. See Andrews and Evans, 2008. I provided a fuller review of medical geography and health geography, and also points of convergence with political ecology studies, in King (2010). Rather than recount that review, my objective here is to highlight key themes that contribute to the social ecology of health framework.

66. See May, 1958; Mayer, 1996.

67. Mayer, 1996, p. 441.

68. Gesler, 2003.

69. Kearns and Moon, 2002, p. 606.

70. Dyck, 1999, p. 247.

71. Kearns and Moon, 2002, p. 611.

72. See Smyth, 2005. Health geography is distinguished not only by its theoretical contributions, but also by its use of diverse methodologies, including participant observation and ethnography. While medical geography has shown a willingness to engage with these approaches, the different methodological and epistemological features of health geography warrant emphasis. Contributing to an emerging health geography, Dyck (1999, p. 245) wrote, "The positivist paradigm that has dominated traditional inquiry, with its focus on geometric space and space as a container of action, is questioned as understandings of space and place are adopted that emphasize their relational, social, and recursive dimensions." Approaches from health geography can challenge the biomedical model by engaging with alternative ways of knowing in order to detail the experiences of people living with disease.

73. Harvey, 1993, p. 7.

74. Massey, 1994, 1999.

75. Moore, 1998.

76. Moore, 1998, p. 353 (his emphasis).

77. Cheng et al., 2003, pp. 89–90.

78. Robbins, 2012.

79. King, 2015.

80. Mayer, 1996, p. 449.

81. See Guthman, 2011; Sultana, 2012; Guthman and Mansfield, 2013; King and Crews, 2013; Jackson and Neely, 2015.

82. King and Crews, 2013.

83. Campbell, 1997.

84. See Marks, 2002; Fassin and Schneider, 2003.

85. Farmer, 1999, p. 79 (his emphasis).

86. Turner, 2009.

87. Robbins and Miller, 2013.

88. Robbins and Miller (2013) explained that water bodies do not need to be large to support mosquito breeding. Buckets or bird baths, for example, are sizable enough to allow the propagation of mosquito populations.

89. King and Crews, 2013.

90. World Health Organization, 2013.

1. See IRIN PlusNews, 2012. SANAC is a collection of government, civil-society, and other stakeholders that is intended to guide national responses to HIV, tuberculosis, and sexually transmitted infections. It replaced the 1992 National AIDS Coordinating Committee of South Africa.

2. Ibid.

3. World Health Organization, UNAIDS, and UNICEF, 2010.

4. Such statistics led, in part, to a 2011 cover story in the *Economist* titled "The End of AIDS?"

5. World Health Organization, UNAIDS, and UNICEF, 2010.

6. See Beaubien, 2013b. Interviewed in the report was Dr. François Venter from the University of Witwatersrand's Reproductive Health and HIV Institute, who called drug shortages the greatest threat to the national HIV program.

7. McNeil, 2015.

8. Centers for Disease Control and Prevention, 2013.

9. See Russell et al., 2007; Swendeman et al., 2009; Kendall and Hill, 2010; McGrath et al., 2014.

10. Beaubien, 2013a.

11. UNAIDS, 2014.

12. In his analysis of disease emergence, Farmer (1999) noted that many studies are at the national level, emphasizing that "The dynamics of disease emergence are not captured in nation-by-nation analyses any more than the diseases are contained by national boundaries, which are themselves emerging entities" (pp. 42–43).

13. Jones, 2005.

14. Saunders (1979, p. 139) reported that in Cape Town, government commissions made recommendations to ensure that Africans and Europeans be separated as far as possible from each other, and that "Africans should either be housed in a compound at the docks or in a general location"—ostensibly to protect whites from an outbreak of the bubonic plague.

15. One of the distinguishing features of how HIV/AIDS was conceptualized by South African national and provincial governmental officials was the use of development terms. Mbeki regularly asserted that HIV should be seen as part of a gamut of diseases that disproportionately affect the poor. Much has been written about the "denialist" position within the South African state at that time. At the center of attention was Mbeki's seeming embrace of the views of a group of scientists that downplayed the link between HIV and AIDS and the viability of

particular treatment regimens. This received international attention with the Treatment Action Campaign's attempts to force the state to provide nevirapine to HIV-positive pregnant mothers, which ultimately led to a showdown in the Constitutional Court that called for the lifting of restrictions on its distribution.

16. Fassin and Schneider, 2003.

17. Chigwedere et al., 2008.

18. Dugger, 2009b, p. A6.

19. Ibid.

20. See Republic of South Africa (2012, p. 12). I use direct quotes from the report to emphasize the discursive framings of the epidemic. The document is strategic in describing the virus as under control and being effectively managed.

21. Beaubien, 2013a.

22. McGrath et al., 2014, p. 303.

23. Kendall and Hill, 2010, pp. 175–176.

24. Rangan and Gilmartin, 2002.

25. Mather, 2000.

26. In King (2005), I pointed out that the process of land allocation was biased toward particular community members, thereby enabling financial elites to benefit from the sugarcane project. The *induna* (local representative of the Matsamo Tribal Authority) and members of his family were given a sugarcane plot, as was one of the local ward councilors. All plots reported by households had male owners. The latter point is particularly problematic because part of the impetus behind the initiative was female empowerment. As a promotional report explained, this was one of seven such projects "aimed at helping previously disadvantaged South Africans, especially women, join more than 50,000 already established small sugar cane growers and tap into one of the country's biggest foreign exchange earners" (African Connexion, 2002, p. 28). During fieldwork in 2001–02, I interviewed a group of ten female farmers who had pressured the Department of Land Affairs to give them a plot farther east of the community. Follow-up fieldwork in 2013 revealed that the Mzinti sugarcane project was unsuccessful and that the fields were no longer in production. Interviews with Mzinti residents suggested that the loan structure didn't provide incentives to produce the sugarcane efficiently. As one of my informants explained, "They ate the profits."

27. This is part of the overarching design of the research project because of the limited number of studies that evaluate the interactions between HIV/AIDS and the natural environment over time. In a comprehensive review of research in Sub-Saharan Africa, Bolton and Talman (2010) argued that local and empirical assessments of these relationships are largely anecdotal, with few studies addressing the environmental and ecosystem impacts over extended periods. They concluded that "snapshot information gives a quick glimpse of issues, but without long-term follow-up the view is likely distorted, especially when dealing

with ecological, health-related, and socio-economic conditions that are in a stake of flux" (p. 27).

28. Acocks, 1953.

29. Republic of South Africa, 2012.

30. Fassin (2007) in particular has documented early HIV/AIDS responses by the national government to reveal how simplistic depictions of AIDS denialism and acceptance, the effectiveness and toxicity of ARVs, and traditional and Western medicine, did not capture the nuance at play within the country at the time.

31. Throughout the book, I use pseudonyms to refer to members of my research team.

32. In a couple of cases the audio recording didn't work effectively, so transcripts of the focus-group interviews were generated by the facilitators, because of their fluency in English and SiSwati.

33. This was a reflection of Thami's personal interest and his having a father who was very knowledgeable about the subject. There are generational differences in engagement with traditional medicine, with older residents being more likely to seek out traditional healers.

34. In a detailed study in South Africa, Posel et al. (2007) drew from focus-group interviews to evaluate perspectives on HIV/AIDS and economic opportunities in the new South Africa. They reported that respondents, presumably referring to AIDS mortality, refer to a "bad" death as being associated with social and cultural causes rather than causes that might be described as biological or physical. As the authors commented, "Indeed, in the main, the [focus group discussions] were far less interested in, and preoccupied by, the immediate problems of bodily illness and the problems of treatment; death was understood and explained through an avowedly social lens—a 'bad' death being first and foremost a symptom of a cultural and moral condition" (p. 141).

35. This underscores part of the intent of the Treatment Action Campaign's slogan "HIV Positive." While reinforcing the possibility of being positive about the virus and having a good quality of life through the use of ARVs, it also emphasized the need to be open about one's status. Compared to reports from other African countries, such as Uganda, individuals in South Africa have been less willing to report their HIV status.

36. Posel et al., 2007.

37. This percentage is averaged between the three villages and obscures variation in reportage of HIV-positive status. The percentage of household heads reporting their status as positive in Mzinti was 24 percent, compared to 14 percent in Schoemansdal and 5 percent in Ntunda. It is also important to emphasize that the 15 percent is a measure of interviewed household heads, and not their entire household. When the survey data were evaluated for all 1,546 individuals within the 327 surveyed households, the percentage reporting HIV-positive status was 6 percent.

38. Kalofonos, 2010.

39. Ibid., p. 364.

40. Food and Agriculture Organization, 2008.

41. The governance framework for the new government was the Reconstruction and Development Programme (RDP), which was drafted by the tripartite alliance of the African National Congress, the South African Communist Party, and the Congress of South African Trade Unions. Levin and Weiner (1997) suggested that although it was accepted by the national conference of the Alliance in January 1994, participation was limited. The RDP was intent on addressing historical inequities and committed the government to investing in the fields of education, democracy and governance, agriculture, business development, health, and housing. One visible sign of this spending in the study area are the government-built houses in the three communities. The Mzinti houses were completed in 2000 and are commonly referred to the as the "RDP homes." As part of the Mbeki presidency, the RDP was largely abandoned in favor of the Growth, Employment and Redistribution Act, which emphasized foreign direct investment and market mechanisms as means to generate economic growth in order to achieve poverty alleviation.

42. Els, 1996.

43. The currency exchange rate was 10 South African rand to 1 U.S. dollar at this point in 2013. This is the exchange rate I use throughout the book.

CHAPTER 3: HISTORICAL SPACES AND
CONTEMPORARY EPIDEMICS

1. Hart (2002) provided a fuller review of the native reserves under British colonialism.

2. Ntsebeza, 2000, p. 287.

3. Ntsebeza, 2000.

4. Ibid.; see also King, 2005.

5. Ntsebeza, 2000.

6. King and McCusker, 2007.

7. Platzky and Walker, 1985.

8. See Pickles and Weiner, 1991; King, 2007; King and McCusker, 2007.

9. King, 2007.

10. Malan and Hattingh, 1976, p. 6.

11. Malan and Hattingh, 1976.

12. Development Bank of Southern Africa, 1987.

13. Malan and Hattingh, 1976.

14. Griffiths and Funnell, 1991.

15. Development Bank of Southern Africa, 1985.

16. KaNgwane was incorporated into the Nkomazi Municipality, which is one of four that constitute the Ehlanzeni District Municipality. Each municipality has a "municipal mayor," who is appointed by the executive council, and another mayor who is elected. The smallest administrative unit is the ward; wards are roughly divided by population size. Each ward has a local councilor. The Mzinti community is divided into two wards, one of them incorporating the village of Ntunda to the south. The ward system was finalized in 2000 and attempts to integrate the various political officials in the country, including councilors, with village and land trusts and the tribal authorities.

17. Levin and Mkhabela, 1997.

18. Niekerk, 1990.

19. Polgreen, 2012.

20. Scoones, 1998, p. 5.

21. Long, 2000, p. 186.

22. In King (2011), I provided a fuller discussion of the livelihood concept.

23. Chambers and Conway, 1992; Chambers, 1997; Leach et al., 1999; Ellis, 2000; de Haan and Zoomers, 2005; Berry, 2009; Ribot, 2009.

24. Robert Putnam's (1994) study of regional development in Italy was foundational to the emergence of social capital. Putnam argued that areas with more effective governments and economic development had "horizontal" social relationships based on trust and shared values. The World Bank's Social Capital Working Group (Woolcock and Narayan, 2000, p. 226) defined social capital as the "norms and networks that enable people to act collectively." Because of its broad application, the concept remains a subject of debate (cf. Woolcock, 1998; Fine, 1999; Bebbington et al., 2006).

25. See Bebbington, 1999; Ellis, 2000; Berry, 2009; Ribot, 2009.

26. Ribot and Peluso, 2003, p. 153.

27. Sen, 1981.

28. Bebbington, 1999, p. 2022.

29. See South Africa Department of Information, 1967; Malan and Hattingh, 1976; Development Bank of Southern Africa, 1985, 1987.

30. South Africa Department of Information, 1967, p. 88.

31. Ibid., p. 86.

32. Development Bank of Southern Africa, 1987, p. 40.

33. South Africa Department of Information, 1967.

34. Development Bank of Southern Africa, 1987, p. 41.

35. King, 2007.

36. In one such study, Shackleton and Shackleton (2000) completed a project in Mpumalanga Province and concluded that five resources—fuelwood, construction wood, edible fruits, edible herbs, and medicinal plants—individually accounted for more than 10 percent of the total value per hectare and together represented over 94 percent of the total value per hectare. On an individual

household level, only fuelwood, edible herbs, and medicinal plants contributed 10 percent or more to the total direct use value. They concluded that "resource harvesting is probably the safety net that allows many households to survive in areas of poor agricultural potential, high human populations and low employment opportunities. These activities should receive the same level of extension support from government agencies that arable and livestock agriculture do" (p. 45).

37. Bhat and Jacobs, 1995; Cocks and Møller, 2002; Ross, 2008; Bishop, 2010; King, 2012.

38. Ardington et al., 2013, p. 1.

39. SouthAfrica.info, 2014.

40. Tapscott, 1995, p. 177.

41. King, 2011.

42. For example, Kale (1995) reported that there were as many as two hundred thousand traditional healers in South Africa, upon whom 80 percent of black South Africans depended for care. It is worth noting that even with a rigorous review of the existing literature, I have never been able to confirm these statistics. Regardless, they are widely cited in later studies.

43. I recognize the dilemma of categorizing these types of doctors as "traditional" while referring to other medical practitioners as "Western" or "modern." While some scholars refer to traditional African medicine as *indigenous medicine* or *cultural medicine*, I use the term *traditional medicine* here to align with its general use within the scholarly literature and field setting. For example, the 2008 Draft Policy from the Department of Health uses "African Traditional Medicine (ATM)" to describe these knowledge systems. My referring to it as "traditional" is not intended to reinforce a hierarchy of health-care options whereby Western clinical medicine is given greater credibility.

44. See King 2004, 2012. This statement is kept intentionally loose, given the complex ways in which disease and health care are perceived by Mzinti residents. It is common for residents to describe certain ailments as traditional maladies that necessitate traditional medicine for treatment. In the focus groups, one man insisted that traditional medicine can be used to get people out of prison if they are caught stealing. Members of a different focus group said they preferred traditional medicine because it can clean your blood and that traditional healers are stronger than clinics. Respondents also spatialized traditional medicine—for example, by describing *muti* from Swaziland as being more powerful. The consequence is that reporting of medical care and treatment should be understood as contingent and complex.

45. King, 2012.

46. This statement so neatly encapsulated the plurality of views around medicinal options that I adopted it as the title of a journal article on the subject.

47. Ashforth, 2005.

48. Stillwaggon, 2006.
49. Laporta et al., 2007.
50. Jones, 2005, p. 425.
51. Thornton, 2008, p. 131.
52. Department of Health, 2008.
53. Ibid., p. 35.

CHAPTER 4: LANDSCAPES OF HIV

1. HIV and AIDS are often written as "HiV" and "Aids" in South African media sources. For consistency, I have changed the acronyms to reflect how they are used in the rest of the book.
2. Sibiya, 2012.
3. Mpumalanga Mirror, 2013.
4. These materials were shared with me by a local AIDS activist working in Mpumalanga in 2006.
5. Sauer, 1925.
6. See Cosgrove, 2003; Mitchell, 2003; Neumann, 2011. In a review of political ecology's contributions to landscape, Neumann (2011) suggested there are theoretical parallels in the fields because landscapes include struggles over identity, multiple and contested meanings of nature, and land rights and use.
7. Leach et al., 1999, p. 239.
8. See Foucault, 1978; Schmitt, 2005.
9. Scott's *Seeing Like a State* (1998) is a masterful analysis of the ambitions of the modern state, suggesting that the most severe instances of state engineering rely on four elements: the administrative ordering of nature and society, High Modernist ideology, an authoritarian state, and a weakened civil society that cannot resist state interventions. In a conclusion that is relevant to the social sciences' burgeoning interest in governmentality, Scott noted that "modern statecraft is largely a project of internal colonization" (p. 82).
10. Brown and Knopp (2010) have argued that biopower is produced through the integration of anatamo-politics and biopolitics. This joining occurs through spatial imaginaries that consider both a politics of the human body and a biopolitics of the population. Using a case study of public health interventions surrounding venereal disease in mid-twentieth-century Seattle, they concluded that social *and* spatial representations are utilized by the state to frame disease and manage people.
11. Nguyen, 2010, pp. 6–7.
12. Biehl, 2011, p. 114.
13. Ibid., p. 115.
14. Ibid., p. 120.

15. Fassin (2010) made a similar point in arguing that the politics of survival challenges a neat duality between the political and the biological. Drawing upon ethnographies of HIV-infected people in South Africa, he concluded (p. 94) that

> powers like the market and the state do act sometimes as if human beings could be reduced to 'mere life,' but democratic forces, including those from within the structure of power, tend to produce alternative strategies that escape this reduction. And people themselves, even under conditions of domination, manage subtle tactics that transform their physical life into a political instrument or a moral resource or an affective expression.

16. Fassin, 2010.

17. Ibid., p. 83 (Fassin framed his essay in response to Jacques Derrida's final interview, which was published in *Le Monde* on August 19, 2004).

18. Scrubb, 2011, p. 2.

19. Ibid.

20. See World Health Organization, 2009; World Health Organization and UNAIDS, 2011.

21. Thien and Del Casino, 2012, p. 1148.

22. Hunter (2010b) asserted that the decline in marriage rates is reworking social relationships in profound ways. He focused on the materiality of everyday sex, whereby boyfriend–girlfriend "gift" relationships are significant in explaining the South African HIV/AIDS epidemic.

23. It should be noted that this protocol does not adhere to the 2013 WHO guidelines. Fieldwork conducted in January 2016 revealed that HIV-positive individuals are not immediately put on ART as recommended by the 2015 WHO guidelines.

24. Jones, 2004.

25. Cited in Republic of South Africa, 2012.

26. This is also reported in the 2012 Human Science Research Council report (Shisana et al., 2014).

27. Republic of South Africa, 2012, p. 53.

28. Human Science Research Council, 2005 (reported in Hunter, 2010a).

29. It bears mentioning that racial categories in post-apartheid South Africa remain restrictive and reinforce apartheid-era classifications. The four categories of white, Indian, coloured, and African remain discursively employed and utilized within national survey instruments. During colonialism and apartheid, "whites" believed to be "pure" descendants of European settlers were separated from the native or African populations within the country. The white population was divided between those of Dutch ancestry, also known as Afrikaners, and those of English heritage. The "coloured" population represents those of mixed ancestry.

30. Republic of South Africa, 2012.

31. Kleinschmidt et al., 2007.

32. Scrubb, 2011, p. 1 (citing Kon and Lackan, 2008).

33. Chopra et al., 2009.

34. Harris et al., 2011.

35. Honda et al., 2015.

36. Els, 1996.

37. van Rooyen, 2016.

38. See Beisel, 2002; Friis, 1998 (cited in Kaschula, 2008).

39. See Melikian et al., 2001; Beisel, 2002.

40. McGarry and Shackleton, 2009.

41. A more alarmist piece in *New Scientist* concluded that affected families are forced to "plunder biodiversity to survive" (Aldhous, 2007, p. 7).

42. Hunter et al., 2007.

CHAPTER 5: HEALTH ECOLOGIES
WITHIN DYNAMIC SYSTEMS

1. Collinge et al., 2008.

2. See Shinn et al., 2014; King et al. 2016. I use the term *secondary floodplain* because it is widely understood; however, some of the Okavango Delta is technically classified as an *occasionally inundated floodplain*.

3. World Health Organization, 2011.

4. Giashuddin et al., 2009.

5. Root and Emch, 2013.

6. World Health Organization, 2011.

7. Ibid.

8. See Paaijmans et al., 2010; Lambrechts et al., 2011; Qi et al., 2013; Smith et al., 2014.

9. Robbins and Miller, 2013.

10. Smith et al., 2014.

11. See McMichael et al., 2007; Piesse and Thirtle, 2009.

12. Ramin and McMichael, 2009 (cited in Smith et al., 2014).

13. Michon et al., 2007.

14. Wilhelmi et al., 2013.

15. Noting the difficulty of defining the term, Harrison and Pearce (2001) referenced a Climate Institute study (Myers and Kent, 1995) that identified *environmental refugees* as those displaced by land shortages, deforestation, soil erosion, desertification, water deficits, extreme weather events, and disease.

16. Centers for Disease Control and Prevention, 2015.

17. World Health Organization, 2010 (cited in Smith et al., 2014).

18. Packard, 2007, p. 25.

19. Alsop, 2007.

20. Alonso et al., 2011.

21. Paaijmans et al., 2010.

22. Shanks et al., 2005.

23. Smith et al., 2014.

24. Ibid., p. 15.

25. Farmer, 1999, p. 39.

26. Beaubien, 2014.

27. Sledge and Mohler, 2013.

28. Carter, 2012.

29. Ibid., p. 4.

30. Packard, 2007.

31. Turshen, 1984, pp. 14–15.

32. Ibid., p. 37.

33. Packard, 2007, p. 11.

34. Boko et al., 2007.

35. IPCC, 2013.

36. Oppenheimer et al., 2014, p. 1056.

37. Ibid.

38. Niang et al., 2014.

39. Murray-Hudson et al., 2006.

40. Andersson et al., 2006, p. 54.

41. See McCarthy et al., 2000; Gumbricht et al., 2001; Wolski et al., 2006.

42. Wolski, 2012.

43. Murray-Hudson et al., 2006.

44. Shinn et al., 2014.

45. As discussed in chapter 3, a livelihood comprises the means through which individuals, families, households, and other social groups meet basic needs and produce material goods and/or income for survival.

46. Mosate, 2010.

47. Etsha 13 is located along the western edge of the Okavango Delta. Similar interviewing was conducted in the villages of Seronga and Mababe in 2012. Seronga is across the Okavango River on the northern extent of the delta, and Mababe is along the eastern edge of the system near Chobe National Park and the Moremi Game Reserve. These three villages were deliberately selected to evaluate the role of location and ecological dynamics within the Okavango Delta.

48. The government and some local residents referred to these encampments as "refugee camps." Regardless of the terminology employed, they illustrate the types of human disruptions that are caused by ecological variability and presage the kind of future investments that will be needed to assist populations displaced as a result of rising sea levels, food shortages, or the breakdown of social systems.

49. Shinn et al. (2014) goes into detail about the intra-village impacts of flooding within Etsha 13. It is worth noting that in some areas the water did recede after the completion of our fieldwork in 2012.

50. Arntzen, 1994.

51. Ringrose et al., 1996.

52. The research project in the Boteti region was supported by the National Science Foundation GSS [RAPID] award no. 0942211, "RAPID: Perceptions of and Adaptation to Extreme Flooding Disturbances in the Okavango Delta, Botswana."

53. In addition to the frequency of this practice within the Okavango Delta as a whole, this is also a consequence of the sampling frame that identified respondents living directly adjacent to the Boteti River. Among those living in other parts of the Boteti region, farther away from the river, dryland farming is more widely practiced. Dryland farming is dependent on precipitation and can involve different crops than those grown by molapo farming. Some respondents were engaging in both dryland and molapo farming and had intricate strategies for balancing these different agricultural types, given the variabilities in precipitation and flooding.

54. Magole and Thapelo, 2005.

55. We did not take people's names in any of the interviews, to ensure that the respondents felt secure in speaking with us. At all times our informants gave us permission to record their responses and to share our findings in governmental reports and other forms of information dissemination. In this interview the eldest daughter volunteered her name, the only caveat being that I share the eventual publication with her upon my next visit to Botswana. I elected to use a pseudonym, as I have done in the other chapters.

56. See Davies, 1991. Lumpy skin disease is caused by a virus that is transmitted by biting insects, including mosquitoes and flies. Tulman et al. (2001) reported that lumpy skin disease virus (LSDV) is a member of the genus *Capripoxvirus,* in the family Poxviridae, and is the etiologic agent of a cattle disease in Africa. Capripoxviruses represent one of eight genera within the chordopoxvirus subfamily of Poxviridae. *Capripoxvirus* currently comprises LSDV, sheeppox virus, and goatpox virus.

57. See Crews (2013) for an expanded discussion of this and its specific relationship to socio-ecological systems.

CHAPTER 6: STATES OF HEALTH

1. These tensions would become more pronounced over the following year. In late 2015, student organizers closed various universities in the country, in protest over governmental proposals to increase tuition rates. Known as the #FeesMust-

Fall campaign, it was one of the most significant protests by youth activists in recent years and served as a direct challenge to the government's ability to provide opportunities to its citizens.

2. Kalofonos, 2010.

3. See Nguyen, 2010; Biehl, 2011.

4. Comprehensive overviews of these fields include Liu et al. (2007) on coupled human and natural systems; Kates et al. (2001) on sustainability science; Ostrom (2008) on socio-ecological systems; and de Sherbinin et al. (2008) on livelihood studies.

5. In attending to HIV/AIDS in particular, I have argued elsewhere that the reification of the shock concept has limitations in understanding the increasingly dynamic features of the epidemic in Sub-Saharan Africa (King, 2013). Focusing on its social and ecological impacts suggests that HIV/AIDS might be better theorized not as a shock but as a health experience that is spatially and temporally dynamic, disproportionate, and dispersed.

6. World Health Organization, 1946 (cited in Grad, 2002, p. 984).

7. Scheper-Hughes (1992) is particularly notable in this regard.

References

Acocks, J. P. H. 1953. *Veld Types of South Africa* (Botanical Survey of South Africa, Memoir 28). Pretoria: Government Printer.

Adams, William M. 2001. *Green Development: Environment and Sustainability in the Third World*, 2nd ed. London: Routledge.

African Connexion. 2002. "Sweat Is Sweet" [promotional report], pp. 28–30. Year 17, first quarter.

Aldhous, Peter. 2007. "The hidden tragedy of Africa's HIV crisis." *New Scientist*, July 11: 6–9.

Alonso, David, Menno J. Bouma, and Mercedes Pascual. 2011. "Epidemic malaria and warmer temperatures in recent decades in an East African highland." *Proceedings of the Royal Society B* 278: 1661–1669.

Alsop, Zoe. 2007. "Malaria returns to Kenya's highlands as temperatures rise." *Lancet* 370: 925–926.

Andersson, Lotta, Julie Wilk, Martin C. Todd, Denis A. Hughes, Anton Earle, Dominic Kniveton, Russel Layberry, and Hubert H. G. Savenije. 2006. "Impact of climate change and development scenarios on flow patterns in the Okavango River." *Journal of Hydrology* 331: 43–57.

Andrews, Gavin J., and Josh Evans. 2008. "Understanding the reproduction of health care: towards geographies in health care work." *Progress in Human Geography* 32: 759–780.

Ardington, Cally, Anne Case, Alicia Menendez, Till Bärninghausen, David Lam, and Murray Leibbrandt. 2013. "Youth unemployment and social protection" (*SALDRU Research Brief,* December). University of Cape Town: SALDRU.

Arntzen, Jaap W. 1994. *Desertification and Possible Solutions in the Mid-Boteti River Area: A Botswana Case Study for the Intergovernmental Convention to Combat Desertification (INCD).* Gaborone: Ministry of Agriculture, Government of Botswana.

Ashforth, Adam. 2005. "*Muthi,* medicine and witchcraft: regulating 'African science' in post-apartheid South Africa?" *Social Dynamics: A Journal of African Studies* 31: 211–242.

Associated Press. 2008a. "Zimbabwe: cholera outbreak kills 294." *New York Times,* November 22: A9.

———. 2008b. "Zimbabwe: Mugabe government is blamed for cholera epidemic." *New York Times,* November 19: A10.

Baer, Hans, and Merrill Singer. 2009. *Global Warming and the Political Ecology of Health: Emerging Crises and Systemic Solutions.* Walnut Creek, California: Left Coast Press.

Barnett, Tony, and Alan Whiteside. 2006. *AIDS in the Twenty-First Century: Disease and Globalization,* 2nd ed. New York: Palgrave Macmillan.

BBC News. 2008. "UK caused cholera, says Zimbabwe." *BBC,* December 12. http://news.bbc.co.uk/2/hi/7780728.stm.

———. 2014. "Liberia to receive Zmapp drug to treat Ebola virus." *BBC,* August 12. www.bbc.com/news/world-africa-28749615.

Bearak, Barry. 2008. "Zimbabwe: anti-cholera measure." *New York Times,* December 2: A16.

Beaubien, Jason. 2013a. "After missteps in HIV care, South Africa finds its way." *All Things Considered.* National Public Radio, August 27. www.npr .org/sections/health-shots/2013/08/27/215734826/after-missteps-in-hiv-care-south-africa-finds-its-way.

———. 2013b. "South Africa weighs starting HIV drug treatment sooner." *Morning Edition.* National Public Radio, July 16. www.npr.org/sections/health-shots/2013/07/16/202381945/ south-africa-weighs-starting-hiv-drug-treatment-sooner.

———. 2014. "Why ending malaria may be more about backhoes than bed nets." *Morning Edition.* National Public Radio, January 3. www.npr.org:sections: health-shots:2014:01:03:257627285:why-ending-malaria-may-be-more-about-backhoes-than-bed-nets.

Bebbington, Anthony. 1999. "Capitals and capabilities: a framework for analyzing peasant viability, rural livelihoods and poverty." *World Development* 27: 2021–2044.

Bebbington, Anthony, Michael Woolcock, Scott Guggenheim, and Elizabeth A. Olson, eds. 2006. *The Search for Empowerment: Social Capital as Idea and Practice at the World Bank.* Bloomfield, Connecticut: Kumarian Press.

Beck, Eckardt C. 1979. "The Love Canal tragedy." *EPA Journal*, January. www2. epa.gov/aboutepa/love-canal-tragedy.

Beisel, William R. 2002. "Nutritionally acquired immune deficiency syndromes." In *Micronutrients and HIV Infection*, edited by Henrik Friis, 23–42. Boca Raton, Florida: CRC Press.

Belluck, Pam. 2009. "New hopes on health care for American Indians." *New York Times*, December 2: A1, A28.

Berry, Sara. 2009. "Property, authority and citizenship: land claims, politics and the dynamics of social division in West Africa." *Development and Change* 40: 23–45.

Bhat, Rama B., and Thomas V. Jacobs. 1995. "Traditional herbal medicine in Transkei." *Journal of Ethnopharmacology* 48(11): 7–12.

Biehl, João. 2011. "When people come first: beyond technical and theoretical quick-fixes in global health." In *Global Political Ecology*, edited by Richard Peet, Paul Robbins, and Michael J. Watts, 100–130. London: Routledge Press.

Bishop, Kristina. 2010. "The nature of medicine in South Africa: the intersection of indigenous and biomedicine." Ph.D. dissertation, University of Arizona.

Boko, Michel, Isabelle Niang, Anthony Nyong, Coleen Vogel, Andrew Githeko, Mahmoud Medany, Balgis Osman-Elasha, Ramadjita Tabo, and Pius Yanda. 2007. "Africa." In *Climate Change 2007: Impacts, Adaptation and Vulnerability. Contribution of Working Group II to the Fourth Assessment Report of the Intergovernmental Panel on Climate Change*, 433–467. Cambridge, UK: Cambridge University Press.

Bolton, Susan, and Anna Talman. 2010. "Interactions between HIV/AIDS and the environment: a review of the evidence and recommendations for next steps." Nairobi, Kenya: IUCN ESARO.

Brown, Michael, and Larry Knopp. 2010. "Between anatamo- and bio-politics: geographies of sexual health in wartime Seattle." *Political Geography* 29: 392–403.

Bullard, Robert D. 2002. "Confronting environmental racism in the twenty-first century." *Global Dialogue* 4: 34–48.

———. 2005. "Neighborhoods 'zoned' for garbage." In *The Quest for Environmental Justice: Human Rights and the Politics of Pollution*, edited by Robert D. Bullard, 43–61. San Francisco: Sierra Club Books.

———. 2007. "25th Anniversary of the Warren County PCB landfill protests: communities of color still on frontline of toxic assaults." *Dissident Voice*, May 29.

Butler, Anthony. 2005. "South Africa's HIV/AIDS policy, 1994–2004: how can it be explained?" *African Affairs* 104: 591–614.

Campbell, Catherine. 1997. "Migrancy, masculine identities and AIDS: the psychosocial context of HIV transmission on the South African gold mines." *Social Science and Medicine* 45: 273–281.

———. 2003: 'Letting Them Die': Why HIV/AIDS Intervention Programmes Fail. Oxford, UK: James Currey.

Carter, Eric D. 2012. *Enemy in the Blood: Malaria, Environment, and Development in Argentina.* Tuscaloosa: University of Alabama Press.

Centers for Disease Control and Prevention. 2005. "Blood lead levels: United States, 1999–2002." *Morbidity and Mortality Weekly Report* 54: 513–516. www.cdc.gov/mmwr/preview/mmwrhtml/mm5420a5.htm.

———. 2013. "About HIV/AIDS." www.cdc.gov/hiv/basics/whatishiv.html.

———. 2015. "Malaria parasites." www.cdc.gov/malaria/about/biology/parasites .html.

Chambers, Robert. 1997. *Whose Reality Counts? Putting the First Last.* London: ITDG.

Chambers, Robert, and Gordon R. Conway. 1992. "Sustainable rural livelihoods: practical concepts for the 21st century." Institute for Development Studies Discussion Paper 296. Brighton, UK: IDS.

Cheng, Antony S., Linda E. Kruger, and Steven E. Daniels. 2003. "'Place' as an integrating concept in natural resource politics: propositions for a social science research agenda." *Society & Natural Resources* 16: 87–104.

Chigwedere, Pride, George R. Seage III, Sofia Gruskin, Tun-Hou Lee, and Max Essex. 2008. "Estimating the lost benefits of antiretroviral drug use in South Africa." *Journal of Acquired Immune Deficiency Syndrome* 49: 410–415.

Chopra, Mickey, Emmanuelle Daviaud, Robert Pattinson, Sharon Fonn, and Joy E. Lawn. 2009. "Saving the lives of South Africa's mothers, babies, and children: can the health system deliver?" *Lancet* 364: 835–846.

Chung, Chanjin, and Samuel L. Myers, Jr. 1999. "Do the poor pay more for food? An analysis of grocery store availability and food price disparities." *Journal of Consumer Affairs* 33: 276–296.

CNN. 2008. "Zimbabwe: West caused cholera crisis." www.cnn.com/2008 /WORLD/africa/12/09/zimbabwe.cholera/index.html?iref = newssearch.

Cocks, Michelle, and Valerie Møller. 2002. "Use of indigenous and indigenised medicines to enhance personal well-being: a South African case study." *Social Science & Medicine* 54: 387–397.

Collinge, Sharon K., Chris Ray, and Jack F. Cully, Jr. 2008. "Effects of disease on keystone species, dominant species, and their communities." In *Infectious Disease Ecology: Effects of Ecosystems on Disease and of Disease on Ecosystems,* edited by Richard S. Ostfeld, Felicia Keesing, and Valerie T. Eviner, 129–144. Princeton, New Jersey: Princeton University Press.

Coovadia, Hoosen, Rachel Jewkes, Peter Barron, David Sanders, and Diane McIntyre. 2009. "The health and health systems of South Africa: historical roots of current health challenges." *Lancet* 374: 817–834.

Cosgrove, Denis. 2003. "Landscape and the European sense of sight— eyeing nature." In *Handbook of Cultural Geography,* edited by Kay M.

Anderson, Mona Domosh, Steve Pile, and Nigel Thrift, 249–268. London: Sage.

Cotterill, Ronald W., and Andrew W. Franklin. 1995. "The urban grocery store gap." Food Marketing Policy Issue Paper 8. Department of Agricultural and Resource Economics, Food Marketing Policy Center, University of Connecticut.

Crews, Kelley A. 2013. "Positioning health in a socio-ecological systems framework." In *Ecologies and Politics of Health*, edited by Brian King and Kelley A. Crews, 15–32. London: Routledge.

Crews, Kelley A., and Brian King. 2013. "Human health at the nexus of ecologies and politics." In *Ecologies and Politics of Health*, edited by Brian King and Kelley A. Crews, 1–12. London: Routledge.

Dai, Dajun. 2011. "Racial/ethnic and socioeconomic disparities in urban green space accessibility: where to intervene?" *Landscape and Urban Planning* 102: 234–244.

Davey, Basiro, and Clive Seale, eds. 1996. *Experiencing and Explaining Disease*. Buckingham, UK: Open University Press.

Davies, F. Glyn. 1991. "Lumpy skin disease of cattle: a growing problem in Africa and the Near East." *World Animal Review* 68(3): 37–42.

Davies, Susanna. 1996. *Adaptable Livelihoods: Coping with Food Insecurity in the Malian Sahel*. London: Macmillan.

de Haan, Leo, and Annelies Zoomers. 2005. "Exploring the frontier of livelihoods research." *Development and Change* 36: 27–47.

Department of Health. 2008. "Draft policy on African traditional medicine for South Africa." Pretoria: Government Gazette.

Dercon, Stefan, John Hoddinott, and Tassew Woldehanna. 2005. "Shocks and consumption in 15 Ethiopian villages, 1999–2004." *Journal of African Economies* 14: 559–585.

de Sherbinin, Alex, Leah K. VanWey, Kendra McSweeney, Rimjhim Aggarwal, Alisson Barbieri, Sabine Henry, Lori M. Hunter, Wayne Twine, and Robert Walker. 2008. "Rural household demographics, livelihoods and the environment." *Global Environmental Change* 18: 38–53.

Development Bank of Southern Africa. 1985. "KaNgwane development information." Sandton, South Africa: Author.

———. 1987. "KaNgwane: introductory economic and social memorandum South Africa. Sandton, South Africa: Author.

de Waal, Alex, and Alan Whiteside. 2003. "New variant famine: AIDS and the food crisis in southern Africa." *Lancet* 362: 1234–1237.

Dionne, Kim Yi. 2014. "Ebola experimental treatment only for the exceptional." *Washington Post*, August 10. www.washingtonpost.com/news/monkey-cage/wp/2014/08/10/ebola-experimental-treatment-only-for-the-exceptional/.

Drimie, Scott. 2003. "HIV/AIDS and land: case studies from Kenya, Lesotho and South Africa." *Development Southern Africa* 20: 647–658.

Dugger, Celia W. 2009a. "Breaking with past, South Africa issues broad AIDS policy." *New York Times,* December 2: A6.

———. 2009b. "South African leader, rejecting predecessor's stance, rallies nation to fight AIDS." *New York Times,* November 1: A20.

Dyck, Isabel. 1999. "Using qualitative methods in medical geography: deconstructive moments in a subdiscipline?" *Professional Geographer* 51: 243–253.

Economist [Editors]. 2011. "The end of AIDS?" *Economist,* June 4: 11.

Ellis, Frank. 2000. *Rural Livelihoods and Diversity in Developing Countries.* Oxford, UK: Oxford University Press.

Els, Herman. 1996. "Game ranching and rural development." In *Game Ranch Management,* edited by J. du P. Bothma, 581–591. Pretoria: Van Schaik.

Environmental Protection Agency. 1994. "What is environmental justice?" Accessed on October 28, 2015. www3.epa.gov/environmentaljustice/.

Faber, Daniel. 1993. *Environment under Fire: Imperialism and the Ecological Crisis in Central America.* New York: Monthly Review Press.

Farmer, Paul. 1999. *Infections and Inequalities: The Modern Plagues.* Berkeley: University of California Press.

———. 2005. *Pathologies of Power: Health, Human Rights, and the New War on the Poor.* Berkeley: University of California Press.

Farmer, Paul, Matthew Basilico, Vanessa Kerry, Madeleine Ballard, Anne Becker, Gene Bukham, Ophelia Dahl, Anne Ellner, Louise Ivers, David Jones, et al. 2013. "Global health priorities for the early twenty-first century." In *Reimagining Global Health: An Introduction,* edited by Paul Farmer, Jim Yong Kim, Arthur Kleinman, and Matthew Basilico, 302–339. Berkeley: University of California Press.

Fassin, Didier. 2007. *When Bodies Remember: Experiences and Politics of AIDS in South Africa.* Berkeley: University of California Press.

———. 2010. "Ethics of survival: a democratic approach to the politics of life." *Humanity* 1: 81–95.

Fassin, Didier, and Helen Schneider. 2003. "The politics of AIDS in South Africa: beyond the controversies." *British Medical Journal* 326: 495–497.

Ferguson, James. 1990. *The Anti-Politics Machine: "Development," Depoliticization and Bureaucratic Power in Lesotho.* New York: Cambridge University Press.

Fine, Ben. 1999. "The developmental state is dead—long live social capital?" *Development and Change* 30: 1–19.

Food and Agriculture Organization. 2008. "An introduction to the basic concepts of food security." www.fao.org/docrep/013/al936e/al936e00.pdf.

Foucault, Michel. 1978. *The History of Sexuality Volume 1: The Will to Knowledge.* London: Penguin.

Friis, Henrik. 1998. "The possible role of micronutrients in HIV infection." *SCN News* 17: 11–12.

Gatrell, Anthony C., and Susan J. Elliott. 2009. *Geographies of Health: An Introduction*, 2nd ed. West Sussex, UK: Wiley-Blackwell.

Gesler, Wilbert. 2003. "Medical geography." In *Geography in America at the Dawn of the 21st Century*, edited by Gary L. Gaile and Cort J. Willmott, 492–502. Oxford, UK: Oxford University Press.

Giashuddin, Sheikh M., Aminur Rahman, Fazlur Rhaman, Saidur Rahman Mashreky, Salim Mahmud Chowdury, Michael Linnan, and Shumona Shafinaz. 2009. "Socioeconomic inequality in child injury in Bangladesh: implication for developing countries." *International Journal for Equity in Health* 8: 7.

Goldman, Michael. 2005. *Imperial Nature: The World Bank and Struggles for Justice in the Age of Globalization*. New Haven, Connecticut: Yale University Press.

Grad, Frank P. 2002. "The preamble of the constitution of the World Health Organization." *Bulletin of the World Health Organization* 80: 981–984.

Griffiths, Ieuan Ll., and D.C. Funnell. 1991. "The abortive Swazi land deal." *African Affairs* 90: 51–64.

Gumbricht, Thomas, Terence McCarthy, and Charles L. Merry. 2001. "The topography of the Okavango Delta, Botswana, and its tectonic and sedimentological implications." *South African Journal of Geology* 104: 243–264.

Guthman, Julie. 2011. *Weighing In: Obesity, Food Justice, and the Limits of Capitalism*. Berkeley: University of California Press.

Guthman, Julie, and Becky Mansfield. 2013. "The implications of environmental epigenetics: a new direction for geographic inquiry on health, space, and nature–society relations." *Progress in Human Geography* 37: 486–504.

Harris, Bronwyn, Jane Goudge, John E. Ataguba, Diane McIntyre, Nonhlanhla Nxumalo, Jikwana Siyabonga, and Matthew Chersich. 2011. "Inequities in access to health care in South Africa." *Journal of Public Health Policy* 32: S102–S123.

Harrison, Jill L. 2011. *Pesticide Drift and the Pursuit of Environmental Justice*. Cambridge, Massachusetts: MIT Press.

Harrison, Paul, and Fred Pearce. 2001. *Atlas of Population and Environment*. American Association for the Advancement of Science (AAAS). Berkeley: University of California Press.

Hart, Gillian. 2002. *Disabling Globalization: Places of Power in Post-Apartheid South Africa*. Berkeley: University of California Press.

Harvey, David. 1993. "From space to place and back again: reflections on the condition of postmodernity." In *Mapping the Futures: Local Cultures and Global Change*, edited by Jon Bird, Barry Curtis, Tim Putnam, George Robertson, and Lisa Tickner, 2–29. New York: Routledge.

Heynen, Nik, Harold A. Perkins, and Parama Roy. 2006. "The political ecology of uneven urban green space: the impact of political economy on race and ethnicity in producing environmental inequality in Milwaukee." *Urban Affairs Review* 42: 3–25.

Hillier, Amy E. 2005. "Residential security maps and neighborhood appraisals: the home owners' loan corporation and the case of Philadelphia." *Social Science History* 29: 207–233.

Honda, Ayako, Mandy Ryan, Robert van Niekerk, and Diane McIntyre. 2015. "Improving the public health sector in South Africa: eliciting public preferences using a discrete choice experiment." *Health Policy and Planning* 30: 600–611.

Human Science Research Council. 2005. *South African National HIV Prevalence, HIV Incidence, Behavior and Communication Survey*. HSRC Press: Cape Town.

Hunter, Lori M., Wayne Twine, and Laura Patterson. 2007. "'Locusts are now our beef': adult mortality and household dietary use of local environmental resources in rural South Africa." *Scandinavian Journal of Public Health* 35(69 Supplement): 165–174.

Hunter, Mark. 2010a. "Beyond the male-migrant: South Africa's long history of health geography and the contemporary AIDS pandemic." *Health & Place* 16: 25–33.

———. 2010b. *Love in the Time of AIDS: Inequality, Gender, and Rights in South Africa*. Bloomington: Indiana University Press.

IPCC. 2013: "Summary for policymakers." In *Climate Change 2013: The Physical Science Basis. Contribution of Working Group I to the Fifth Assessment Report of the Intergovernmental Panel on Climate Change*. New York: Cambridge University Press.

IRIN PlusNews. 2012. "South Africa: revamped AIDS council makes its debut." *IRIN*, October 9. www.irinnews.org/report/96492/south-africa-revamped-aids-council-makes-its-debut.

Jackson, Paul, and Abigail H. Neely. 2015. "Triangulating health: toward a practice of a political ecology of health." *Progress in Human Geography* 39: 47–64.

Jones, Peris S. 2004. "When 'development' devastates: donor discourses, access to HIV/AIDS treatment in Africa and rethinking the landscape of development." *Third World Quarterly* 25: 385–404.

———. 2005. "'A test of governance': rights-based struggles and the politics of HIV/AIDS policy in South Africa." *Political Geography* 24: 419–447.

Kale, Rajendra. 1995. "Traditional healers in South Africa: a parallel health care system." *British Medical Journal* 310: 1182–1185.

Kalofonos, Ippolytos A. 2010. "'All I eat is ARVs': the paradox of AIDS treatment interventions in central Mozambique." *Medical Anthropology Quarterly* 24: 363–380.

Kaschula, Sarah A. 2008. "Wild foods and household food security responses to AIDS: evidence from South Africa." *Population and Environment* 29: 162–185.

Kates, Robert W., William C. Clark, Robert Corell, J. Michael Hall, Carlo C. Jaeger, Ian Lowe, James J. McCarthy, Hans J. Schellnhuber, Bert Bolin, Nancy M. Dickson, et al. 2001. "Environment and development: sustainability science." *Science* 292: 641–642.

Kaufman, Phil R. 1999. "Rural poor have less access to supermarkets, large grocery stores." *Rural Development Perspectives* 13(3): 19–25.

Kearns, Robin, and Graham Moon. 2002. "From medical to health geography: novelty, place and theory after a decade of change." *Progress in Human Geography* 26: 605–625.

Kendall, Carl, and Zelee Hill. 2010. "Chronicity and AIDS in three South African communities." In *Chronic Conditions, Fluid States: Chronicity and the Anthropology of Illness,* edited by Lenore Manderson and Carolyn Smith-Morries, 175–194. New Brunswick, New Jersey: Rutgers University Press.

Kgathi, Donald L., Barbara N. Ngwenya, and Julie Wilk. 2007. "Shocks and rural livelihoods in the Okavango Delta, Botswana." *Development Southern Africa* 24: 289–308.

King, Brian. 2004. "In the shadow of Kruger: community conservation and environmental resource access in the former KaNgwane Homeland, South Africa." Ph.D. dissertation, University of Colorado at Boulder.

———. 2005. "Spaces of change: tribal authorities in the former KaNgwane homeland, South Africa." *Area* 37: 64–72.

———. 2007. "Developing KaNgwane: geographies of segregation and integration in the new South Africa." *Geographical Journal* 173: 13–25.

———. 2010. "Political ecologies of health." *Progress in Human Geography* 34: 38–55.

———. 2011. "Spatialising livelihoods: Resource access and livelihood spaces in South Africa." *Transactions of the Institute of British Geographers* 36: 297–313.

———. 2012. "'We pray at the church in the day and visit the *sangomas* at night': Health discourses and traditional medicine in rural South Africa." *Annals of the Association of American Geographers* 102: 1173–1181.

———. 2013. "Disease as shock, HIV/AIDS as experience: coupling social and ecological responses in Sub-Saharan Africa." In *Ecologies and Politics of Health* edited by Brian King and Kelley A. Crews, 260–279. London: Routledge.

———. 2015. "Political ecologies of disease and health." In *The Routledge Handbook of Political Ecology,* edited by Tom Perreault, Gavin Bridge, and James McCarthy, 343–353. London: Routledge.

King, Brian, and Kelley A. Crews, eds. 2013. *Ecologies and Politics of Health.* London: Routledge.

King, Brian, and Brent McCusker. 2007. "Environment and development in the former South African Bantustans." *Geographical Journal* 173: 6–12.

King, Brian, Jamie E. Shinn, Kelley A. Crews, and Kenneth R. Young. 2016. Fluid waters and rigid livelihoods in the Okavango Delta of Botswana. *Land* 5(2): 1–16.

Kleinschmidt, Immo, Audrey Pettifor, Natashia Morris, Catherine MacPhail, and Helen Rees. 2007. "Geographic distribution of human immunodeficiency virus in South Africa." *American Journal of Tropical Medicine and Hygiene* 77: 1163–1169.

Kon, Zelda R., and Nuha Lackan. 2008. "Ethnic disparities in access to care in post-apartheid South Africa." *American Journal of Public Health* 98: 2272–2277.

Krieger, Nancy. 1994. "Epidemiology and the web of causation: has anyone seen the spider?" *Social Science & Medicine* 39: 887–903.

Lalonde, Marc. 1974. "A new perspective on the health of Canadians: a working document." Ottawa: Minister of Supply and Services.

Lambrechts, Louis, Krijn P. Paaijmans, Thanyalak Fansiri, Lauren B. Carrington, Laura D. Kramer, Matthew B. Thomas, and Thomas W. Scott. 2011. "Impact of daily temperature fluctuations on Dengue virus transmission by *Aedes aegypti*." *Proceedings of the National Academy of Sciences USA* 108: 7460–7465.

Laporta, Olga, Laura Pérez-Fons, Ricardo Mallavia, Nuria Caturla, and Vicente Micol. 2007. "Isolation, characterization and antioxidant capacity assessment of the bioactive compounds derived from *Hyposix rooperi* corm extract (African potato)." *Food Chemistry* 101: 1425–1437.

Leach, Melissa, Robin Mearns, and Ian Scoones. 1999. "Environmental entitlements: dynamics and institutions in community-based natural resource management." *World Development* 27: 225–247.

Lerner, Steve. 2010. *Sacrifice Zones: The Front Lines of Toxic Chemical Exposure in the United States.* Cambridge, Massachusetts: MIT Press.

Levin, Richard, and Sam Mkhabela. 1997. "The chieftaincy, land allocation, and democracy." In *No More Tears: Struggles for Land in Mpumalanga, South Africa*, edited by Richard Levin and Daniel Weiner, 153–173. Trenton, New Jersey: Africa World Press.

Levin, Richard, and Daniel Weiner, eds. 1997. *No More Tears: Struggles for Land in Mpumalanga, South Africa.* Trenton, New Jersey: Africa World Press.

Li, Tania Murray. 2007. *The Will to Improve: Governmentality, Development, and the Practice of Politics.* Durham, North Carolina: Duke University Press.

Liu, Jianguo, Thomas Dietz, Stephen R. Carpenter, Marina Alberti, Carl Folke, Emilio Moran, Alice N. Pell, Peter Deadman, Timothy Kratz, Jane Lubchenco, et al. 2007. "Complexity of coupled human and natural systems." *Science* 317: 1513–1516.

Long, Norman. 2000. "Exploring local/global transformations: a view from anthropology." In *Anthropology, Development, and Modernities: Exploring Discourses, Counter-tendencies and Violence,* edited by Alberto Arce and Norman Long, 184–201. London: Routledge.

Love, Roy. 2004. "HIV/AIDS in Africa: links, livelihoods, and legacies." *Review of African Political Economy* 31: 639–648.

Magole, Lapologang, and Kebonyemodia Thapelo. 2005. "The impact of extreme flooding of the Okavango River on the livelihood of the *molapo* farming community of Tubu village, Ngamiland Sub-district, Botswana." *Botswana Notes and Records* 37: 125–137.

Makgoba, Malegapur W. 2002. "Politics, the media and science in HIV/AIDS: the perils of pseudoscience." *Vaccine* 20: 1899–1904.

Malan, Theo, and P. S. Hattingh. 1976. *Black Homelands in South Africa.* Pretoria: Africa Institute of South Africa.

Mansfield, Becky. 2008. "Health as a nature–society question." *Environment and Planning A* 40: 1015–1019.

Marks, Shula. 2002: "An epidemic waiting to happen? The spread of HIV/AIDS in South Africa in social and historical perspective." *African Studies* 61: 13–26.

Massey, Doreen. 1994. *Space, Place and Gender.* Minneapolis: University of Minnesota Press.

———. 1999. "Space-time, 'science,' and the relationship between physical geography and human geography." *Transactions of the Institute of British Geographers* 24: 261–276.

Mather, Charles. 2000. "Foreign migrants in export agriculture: Mozambican labour in the Mpumalanga lowveld, South Africa." *Tijdschrift voor Economische en Sociale Geografie* 91: 426–436.

May, Jacques Meyer. 1958. *The Ecology of Human Disease.* New York: MD Publications.

Mayer, Jonathan D. 1996. "The political ecology of disease as one new focus for medical geography." *Progress in Human Geography* 20: 441–456.

Mayer, Kenneth H., and H. F. Pizer. 2008. "Introduction: what constitutes the social ecology of infectious diseases?" In *The Social Ecology of Infectious Diseases,* edited by Kenneth H. Mayer and H. F. Pizer, 1–16. Burlington, Massachusetts: Elsevier.

McCarthy, Terence S., Gordon R. J. Cooper, P. D. Tyson, and William N. Ellery. 2000. "Seasonal flooding in the Okavango Delta, Botswana—recent history and future prospects." *South African Journal of Science* 96: 25–33.

McGarry, Dylan K., and Charles M. Shackleton. 2009. "Comment: is HIV/AIDS jeopardizing biodiversity?" *Environmental Conservation* 36: 5–7.

McGrath, Janet W., Margaret S. Winchester, David Kaawa-Mafigiri, Eddy Walakira, Florence Namutiibwa, Judith Birungi, George Ssendegye, Amina Nalwoga, Emily Kyarikunda, Sheila Kisakye, et al. 2014. "Challenging the paradigm: anthropological perspectives on HIV as a chronic disease." *Medical Anthropology* 33: 303–317.

McLeod, Kari S. 2000. "Our sense of Snow: the myth of John Snow in medical geography." *Social Science & Medicine* 50: 923–935.

McMichael, Anthony J., John W. Powles, Colin D. Butler, and Ricardo Uauy. 2007. "Food, livestock production, energy, climate change, and health." *Lancet* 370: 1253–1263.

McNeil, Donald G., Jr. 2015. "World health group proposes extending H.I.V. therapy's reach." *New York Times,* October 1: A8.

Melikian, George, Francis Mmiro, Christopher Ndugwa, Robert Perry, J. Brooks Jackson, Elizabeth Garrett, James Tielsch, and Richard D. Semba. 2001. "Relation of vitamin A and carotenoid status to growth failure and mortality among Ugandan infants with human immunodeficiency virus." *Nutrition* 17: 567–572.

Michon, Pascal, Jennifer L. Cole-Tabioan, Elijiah Dabod, Sonja Schoepflin, Jennifer Igu, Melinda Susapu, Nandao Tarongka, Peter A. Zimmerman, John C. Reeder, James G. Beeson, et al. 2007. "The risk of malarial infections and disease in Papua New Guinean children." *American Journal of Tropical Medicine and Hygiene* 76: 997–1008.

Mishler, Elliot G., Lorna R. Amarasingham, Stuart T. Hauser, Ramsay Liem, Samuel D. Osherson, and Nancy E. Waxler. 1981. *Social Contexts of Health, Illness, and Patient Care.* New York: Cambridge University Press.

Mitchell, Don. 2003. "Cultural landscapes: just landscapes or landscapes of justice?" *Progress in Human Geography* 27: 787–796.

Mitchell, Timothy. 2002. *Rule of Experts: Egypt, Techno-Politics, Modernity.* Berkeley: University of California Press.

Moore, Donald. 1998. "Subaltern struggles and the politics of place: remapping resistance in Zimbabwe's eastern highlands." *Cultural Anthropology* 13: 344–381.

Mosate, Masego. 2010. "Managing Botswana's floods." Republic of Botswana National Disaster Management Office, June 18. www.gov.bw/en/Citizens /Topics/Citizen-News/Managing-Botswanas-Floods/.

Mpumalanga Mirror. 2013. "Community to benefit from new campaign." *Mpumalanga Mirror,* June 11: 3.

Murphree, Marshall W. 1990. *Decentralizing the Proprietorship of Wildlife Resources in Zimbabwe's Communal Lands.* Harare: Centre for Applied Social Sciences, University of Zimbabwe.

Murphy, Laura, Paul Harvey, and Eva Silvestre. 2005. "How do we know what we know about the impact of AIDS on food and livelihood insecurity? A review of empirical research from rural sub Saharan Africa." *Human Organization* 64: 265–274.

Murray, Christopher J. L., and Alan D. Lopez. 1997. "Alternative projections of mortality and disability by cause 1990–2020: Global Burden of Disease study." *Lancet* 349: 1498–1504.

Murray-Hudson, Michael, Piotr Wolski, and Susan Ringrose. 2006. "Scenarios of the impact of local and upstream changes in climate and water use on hydro-ecology in the Okavango Delta, Botswana." *Journal of Hydrology* 331: 73–84.

Myers, Norman, and Jennifer Kent. 1995. *Environmental Exodus: An Emergent Crisis in the Global Arena*. Washington, DC: Climate Institute.

Nash, Linda. 2006. *Inescapable Ecologies: A History of Environment, Disease, and Knowledge*. Berkeley: University of California Press.

Nattrass, Nicoli. 2004. *The Moral Economy of AIDS in South Africa*. Cambridge, UK: Cambridge University Press.

Negin, Joel. 2005. "Assessing the impact of HIV/AIDS on economic growth and rural agriculture in Africa." *Journal of International Affairs* 58: 267–281.

Neighmond, Patti. 2014. "It takes more than a produce aisle to refresh a food desert." *Morning Edition*. National Public Radio, February 10. www.npr.org /blogs/thesalt/2014/02/10/273046077/ takes-more-than-a-produce-aisle-to-refresh-a-food-desert.

Neumann, Roderick P. 2011. "Political ecology III: theorizing landscape." *Progress in Human Geography* 35: 843–850.

New York State Department of Health. 2009. "Love Canal follow-up health study cancer study community report." www.health.ny.gov/environmental /investigations/love_canal/cancer_study_community_report.htm.

N'Goran, E. K., S. Diabate, J. Utzinger, and B. Sellin. 1997. "Changes in human schistosomiasis levels after the construction of two large hydroelectric dams in central Côte d'Ivoire." *Bulletin of the World Health Organization* 75: 541–545.

Nguyen, Vinh-Kim. 2010. *The Republic of Therapy: Triage and Sovereignty in West Africa's Time of AIDS*. Durham, North Carolina: Duke University Press.

Niang, Isabelle, Oliver C. Ruppel, Mohamed A. Abdrabo, Ama Essel, Christopher Lennard, Jonathan Padgham, and Penny Urquhart. 2014. "Africa." In *Climate Change 2014: Impacts, Adaptation, and Vulnerability. Part B: Regional Aspects. Contribution of Working Group II to the Fifth Assessment Report of the Intergovernmental Panel on Climate Change*, 1199–1265. New York: Cambridge University Press.

Niekerk, P. V. 1990. "Bantustan politics." *New Ground* 1(1): 12–14.

Ntsebeza, Lungisile. 2000. "Traditional authorities, local government and land rights." In *At the Crossroads: Land and Agrarian Reform in South Africa into the 21st Century: Papers from a Conference Held at Alpha Training Centre, Broederstroom, Pretoria, South Africa, 26-28 July 1999*, edited by Ben Cousins, 280–305. Cape Town: Programme for Land and Agrarian Studies, School of Government, University of the Western Cape and National Land Committee.

Oppenheimer, Michael, Maximilaiano Campos, Rachel Warren, Joern Birkmann, George Luber, Brian O'Neill, and Kiyoshi Takahashi. 2014. "Emergent risks and key vulnerabilities." In *Climate Change 2014: Impacts, Adaptation, and Vulnerability. Part A: Global and Sectoral Aspects. Contribution of Working Group II to the Fifth Assessment Report of the Intergovernmental Panel on Climate Change*, 1039–1099. New York: Cambridge University Press.

Ostrom, Elinor. 2008. "Frameworks and theories of environmental change." *Global Environmental Change* 18: 249–252.

Paaijmans, Krijn P., Simon Blanford, Andrew S. Bell, Justine I. Blanford, Andrew F. Read, and Matthew B. Thomas. 2010. "Influence of climate on malaria transmission depends on daily temperature variation." *Proceedings of the National Academy of Sciences USA* 107: 15135–15139.

Packard, Randall M. 2007. *The Making of a Tropical Disease: A Short History of Malaria*. Baltimore, Maryland: Johns Hopkins University Press.

Parkhurst, Justin O. 2004. "The political environment of HIV: lessons from a comparison of Uganda and South Africa." *Social Science & Medicine* 59: 1913–1924.

Pickles, John, and Daniel Weiner. 1991. "Rural and regional restructuring of apartheid: ideology, development policy, and the competition for space." *Antipode* 23: 2–32.

Piesse, Jenifer, and Colin Thirtle. 2009. "Three bubbles and a panic: an explanatory review of recent food commodity price events." *Food Policy* 34: 119–129.

Platzky, Laurine, and Cherryl Walker. 1985. *The Surplus People: Forced Removals in South Africa*. Johannesburg: Raven Press.

Polgreen, Lydia. 2012. "South Africa debates law to support tribal courts." *New York Times*, June 16.

Posel, Deborah, Kathleen Kahn, and Liz Walker. 2007. "Living with death in a time of AIDS: a rural South African case study." *Scandinavian Journal of Public Health* 35(69 Supplement): 138–146.

Prohaska, Thomas. 2014. "15 new lawsuits filed over Love Canal work." *Buffalo News*, February 14.

Putnam, Robert D. [with Robert Leonardi and Raffaella Y. Nanetti]. 1994. *Making Democracy Work: Civic Traditions in Modern Italy*. Princeton, New Jersey: Princeton University Press.

Qi, Jiaguo, Lindsay P. Campbell, Jenni Van Ravensway, Andrew O. Finley, Richard W. Merritt, and M. Eric Benbow. 2013. "Buruli ulcer disease: the unknown environmental and social ecology of a bacterial pathogen." In *Ecologies and Politics of Health,* edited by Brian King and Kelley A. Crews, 75–97. London: Routledge.

Quammen, David. 2012. *Spillover: Animal Infections and the Next Human Pandemic.* New York: W.W. Norton.

Ramin, Brodie M., and Anthony J. McMichael. 2009. "Climate change and health in sub-Saharan Africa: a case-based perspective." *EcoHealth* 6: 52–57.

Rangan, Haripriya, and Mary Gilmartin. 2002. "Gender, traditional authority, and the politics of rural reform in South Africa." *Development and Change* 33: 633–658.

Republic of South Africa. 2012. "Global AIDS Response Progress Report 2012." http://www.unaids.org/sites/default/files/country/documents //file,69086,es..pdf.

Ribot, Jesse C. 2009. "Authority over forests: empowerment and subordination in Senegal's democratic decentralization." *Development and Change* 40: 105–129.

Ribot, Jesse C., and Nancy Lee Peluso. 2003. "A theory of access." *Rural Sociology* 68: 153–181.

Ringrose, Susan, Raban Chanda, Musisi Nkambwe, and Francis Sefe. 1996. "Environmental change in the mid-Boteti area of north-central Botswana: biophysical processes and human perceptions." *Environmental Management* 20: 397–410.

Robbins, Paul. 2012. *Political Ecology: A Critical Introduction,* 2nd ed. Malden, Massachusetts: Wiley-Blackwell.

Robbins, Paul, and Jacob C. Miller. 2013. "The mosquito state: how technology, capital, and state practice mediate the ecologies of public health." In *Ecologies and Politics of Health,* edited by Brian King and Kelley A. Crews, 196–215. London: Routledge.

Root, Elisabeth D., and Michael Emch. 2013. "The ecology of injuries in Matlab, Bangladesh." In *Ecologies and Politics of Health,* edited by Brian King and Kelley A. Crews, 98–117. London: Routledge.

Ross, Eleanor. 2008. "Traditional healing in South Africa: ethical implications for social work." *Social Work in Health Care* 46(2): 15–33.

Rostow, Walter W. 1960. *The Stages of Economic Growth: A Non-Communist Manifesto.* Cambridge, UK: Cambridge University Press.

Russell, Steven, Janet Seeley, Enoch Ezati, Nafuna Wamai, Willy Were, and Rebecca Bunnell. 2007. "Coming back from the dead: living with HIV as a chronic condition in rural Africa." *Health Policy and Planning* 22: 344–347.

Sauer, Carl O. 1925. *The Morphology of Landscape.* Berkeley: University of California Press.

Saunders, Christopher. 1979. "From Ndabeni to Langa." *Studies in the History of Cape Town* 1: 167–204.

Scheper-Hughes, Nancy. 1992. *Death without Weeping: The Violence of Everyday Life in Brazil.* Berkeley: University of California Press.

Schmitt, Carl. 2005. *Political Theory: Four Chapters on the Concept of Sovereignty,* translated by George Schwab. Chicago, Illinois: University of Chicago Press.

Scoones, Ian. 1998. "Sustainable rural livelihoods: a framework for analysis." IDS Working Paper 72. Brighton, UK: Institute for Development Studies.

Scott, James C. 1998. *Seeing Like a State: How Certain Schemes to Improve the Human Condition Have Failed.* New Haven, Connecticut: Yale University Press.

Scrubb, Victoria. 2011. "Political systems and health inequity: connecting apartheid policies to the HIV/AIDS epidemic in South Africa." *Journal of Global Health,* April 1. www.ghjournal.org/political-systems-and-health-inequity-connecting-apartheid-policies-to-the-hivaids-epidemic-in-south-africa/.

Sen, Amartya. 1981. *Poverty and Famines: An Essay on Entitlement and Deprivation.* New York: Oxford University Press.

———. 1999. *Development as Freedom.* New York: Knopf.

Shackleton, Charles M., and Sheona E. Shackleton. 2000. "Direct use values of secondary resources harvested from communal savannas in the Bushbuckridge lowveld, South Africa." *Journal of Tropical Forest Products* 6: 28–47.

Shanks, G. Dennis, Simon I. Hay, Judy A. Omumbo, and Robert W. Snow. 2005. "Malaria in Kenya's western highlands." *Emerging Infectious Diseases* 11: 1425–1432.

Shinn, Jamie E., Brian King, Kenneth R. Young, and Kelley A. Crews. 2014. "Variable adaptations: micro-politics of environmental displacement in the Okavango Delta, Botswana." *Geoforum* 57: 21–29.

Shisana, Olive, Thomas Rehle, Leickness Simbayi, Khangelani Zuma, Sean Jooste, N. Zungu, Demetre Labadarios, Dorina Onoya, et al. 2014. *South African National HIV Prevalence, Incidence and Behaviour Survey, 2012.* Cape Town: HSRC Press.

Sibiya, D. 2012. "Nkomazi excels in the fight against HIV/Aids." *Corridor Gazette,* November 29: 6.

Singer, Merrill. 1996. "A dose of drugs, a touch of violence, a case of AIDS: conceptualizing the SAVA syndemic." *Free Inquiry in Creative Sociology* 24: 99–110.

Sledge, Daniel, and George Mohler. 2013. "Eliminating malaria in the American South: an analysis of the decline of malaria in 1930s Alabama." *American Journal of Public Health* 103: 1381–1392.

Smith, David. 2009. "New cholera outbreak in Zimbabwe." *Guardian,* August 24. www.guardian.co.uk/world/2009/aug/24/zimbabwe-new-cholera-outbreak.

Smith, Kirk R., Alistair Woodward, D. Campbell-Lendrum, Dave D. Chadee, Yasushi Honda, Qiyong Liu, Jane Olwoch, Boris Revich, and Rainer Sauer-born. 2014. "Human health: impacts, adaptation, and co-benefits." In *Climate Change 2014: Impacts, Adaptation, and Vulnerability. Part A: Global and Sectoral Aspects. Contribution of Working Group II to the Fifth Assessment Report of the Intergovernmental Panel on Climate Change,* 709–754. New York: Cambridge University Press.

Smyth, Fiona. 2005. "Medical geography: therapeutic places, spaces and networks." *Progress in Human Geography* 29: 488–495.

South Africa Department of Information. 1967. *The Progress of the Bantu Peoples towards Nationhood.* Pretoria: Department of Information.

SouthAfrica.info. 2014. "Social grants reach almost one-third of South Africans." June 19. www.southafrica.info/about/social/grants-190614.htm#.Vkm25oSC7Xk.

South Africa Information Service. 1973. *Progress through Separate Development: South Africa in Peaceful Transition,* 4th ed. New York: Information Service of South Africa.

Stillwaggon, Eileen. 2006. *AIDS and the Ecology of Poverty.* New York: Oxford University Press.

Sultana, Farhana. 2012. "Producing contaminated citizens: toward a nature-society geography of health and well-being." *Annals of the Association of American Geographers* 102: 1165–1172.

Swendeman, Dallas, Barbara L. Ingram, and Mary Jane Rotheram-Borus. 2009. "Common elements in self-management of HIV and other chronic illnesses: an integrative framework." *AIDS Care* 21: 1321–1334.

Tapscott, Chris. 1995. "Changing discourses of development in South Africa." In *Power of Development,* edited by Jonathan Crush, 176–191. London: Routledge.

Thien, Deborah, and Vincent J. Del Casino, Jr. 2012. "(Un)healthy men, masculinities, and the geographies of health." *Annals of the Association of American Geographers* 102: 1146–1156.

Thornton, Robert. 2008. *Unimagined Community: Sex, Networks, and AIDS in Uganda and South Africa.* Berkeley: University of California Press.

Tulman, E. R., C. L. Afonso, Z. Lu, L. Zsak, G. F. Kutish, and D. L. Rock. 2001. "Genome of lumpy skin disease virus." *Journal of Virology* 75: 7122–7130.

Turner, Matthew D. 2009. "Ecology: natural and political." In *A Companion to Environmental Geography,* edited by Noel Castree, David Demeritt, Diana Liverman, and Bruce Rhoads, 181–197. Malden, Massachusetts: Blackwell.

Turshen, Meredeth. 1977. "The political ecology of disease." *Review of Radical Political Economics* 9: 45–60.

———. 1984. *The Political Ecology of Disease in Tanzania*. New Brunswick, New Jersey: Rutgers University Press.

UNAIDS. 2014. "South Africa: HIV and AIDS estimates (2014)." www.unaids .org/en/regionscountries/countries/southafrica.

United Nations. 2008. "Millennium Development Goals Indicators: The official United Nations site for the MDG Indicators." http://mdgs.un.org/unsd/mdg /Metadata.aspx?IndicatorId = 0&SeriesId = 711.

———. 2013. "Goal 7: Ensure environmental sustainability." *Millennium Development Goals*. www.un.org/millenniumgoals/environ.shtml.

United Nations News Centre. 2015. "UN health agency warns Ebola outbreak in West Africa has 'a very nasty sting in its tail.'" September 9. www.un.org /apps/news/story.asp?NewsID = 51837#.VfKpVHiC7Xk.

van der Vliet, Virginia. 2004. "South Africa divided against AIDS: a crisis of leadership." In *AIDS and South Africa: The Social Expression of a Pandemic*, edited by Kyle D. Kauffman and David L. Lindauer, 48–96. New York: Palgrave MacMillan.

van Rooyen, Fanie. 2016. "In the grip of a monster El Niño." *Popular Mechanics*, March 16: 44–47.

Wade, Derick T., and Peter W. Halligan. 2004. "Do biomedical models of illness make for good healthcare systems?" *British Medical Journal* 329: 1398.

Wainwright, Joel. 2008. *Decolonizing Development: Colonial Power and the Maya*. Malden, Massachusetts: Blackwell.

Walker, Renee E., Christopher R. Keane, and Jessica G. Burke. 2010. "Disparities and access to healthy food in the United States: a review of food deserts literature." *Health & Place* 16: 876–884.

Webb, Penny, and Chris Bain. 2011. *Essential Epidemiology: An Introduction for Students and Health Professionals*, 2nd ed. New York: Cambridge University Press.

Wilhelmi, Olga, Alex de Sherbinin, and Mary Hayden. 2013. "Exposure to heat stress in urban environments." In *Ecologies and Politics of Health*, edited by Brian King and Kelley A. Crews, 219–238. London: Routledge.

Wines, Michael. 2006. "AIDS cited in the climb in South Africa's death rate." *New York Times*, September 7: A6.

Wolski, Piotr. 2012. "Water discharge at Mohembo." *Okavango Delta Monitoring and Forecasting*. Okavango Research Institute. http://168.167.30.198 /ori/monitoring/water/index.php.

Wolski, Piotr, Hubert H. G. Savenije, Michael Murray-Hudson, and Thomas Gumbricht. 2006. "Modelling of the flooding in the Okavango Delta, Botswana, using a hybrid reservoir–GIS model." *Journal of Hydrology* 331: 58–72.

Woolcock, Michael. 1998. "Social capital and economic development: towards a theoretical synthesis and policy framework." *Theory and Society* 27: 151–208.

Woolcock, Michael, and Deepa Narayan. 2000. "Social capital: implications for development theory, research, and policy." *World Bank Research Observer* 15: 225–249.

World Health Organization. 1946. *World Health Organization Constitution.* Adopted by the International Health Conference, New York, June 19–July 22.

———. 1986. "The Ottawa Charter for Health Promotion." http://www.who.int /healthpromotion/conferences/previous/ottawa/en/.

———. 2009. *Traditional Male Circumcision among Young People: A Public Health Perspective in the Context of HIV Prevention.* Geneva: WHO Press.

———. 2010. *World Malaria Report 2010.* Geneva. Switzerland: WHO Press.

———. 2011. "WHO Public Health & Environment Global Strategy Overview." September 26. www.who.int/phe/publications/PHE_2011_global_strategy_ overview_2011.pdf.

———. 2013. "Environmental health." www.who.int/topics/environmental_ health/en/.

World Health Organization and UNAIDS. 2011. *Progress in Scale-up of Male Circumcision for HIV Prevention in Eastern and Southern Africa: Focus on Service Delivery.* Geneva. Switzerland: WHO Press.

World Health Organization, UNAIDS, and UNICEF. 2010. "Towards universal access: scaling up priority HIV/AIDS interventions in the health sector" (Progress Report). Geneva: WHO HIV/AIDS Department.

Index

Abdullah, Fareed, 51
Adams, William, 33
adaptive capacity, 145
Aedes aegypti, 12
African American population, landfill sitings and, 35–37
African National Congress (ANC), 3, 86, 98; Bantustan traditional authorities and, 86; corruption charges, 164; investment in health-care facilities, 126; national elections (2014), 163–64
African potato, 9, 97–98
Afton community (North Carolina), landfill, 35–36
agricultural extension group (Mzinti), 163
agriculture, 77–79, 89, 90; biofuel demand and, 128; climate change and, 137, 138, 139, 145; dryland farming, 13, 89, 190n53; for export, 128; fertilizers and pesticides, 32; floodplain (molapo) farming, 13, 153, 190n53; gardens for food, 75, 77–78, 138, 165; infrastructure for, 45; land dispossessions, 90, 131; land quality in Bantustans, 85, 89; land used for other purposes, 78; natural environment and, 133; sugarcane, 60–61, 128, 129, 163, 181n26; water availability and, 13; weather and precipitation

and, 132, 138. *See also* food production; food security
AIDS (acquired immunodeficiency syndrome): as a "bad death," 182n34; basics about, 54–56; biomedical model, 24; cause of, 24; death rates in South Africa, 57; deaths attributed to associated illnesses, xii, 6, 9; deaths attributed to poverty, 9; Global AIDS Response Progress Report (2012), 125; as "hidden in plain sight," 6–7, 11, 64; "In South Africa, no one dies of AIDS" statement, 11; provincial AIDS councils, 52; SANAC (South African National AIDS Council), 51, 79; symptoms, 55; women's vulnerability to, 45. *See also* HIV; HIV/AIDS
AIDS policy: under Mbeki, Thabo, xii-xiii, 8, 56, 64, 72, 180–81n15; under Zuma, Jacob, xii, 10–11, 51, 57–58
AIDS treatment. *See* antiretroviral drugs (ARVs); antiretroviral therapy (ART)
air pollution, 135, 136
Alsop, Zoe, 140
American Indians, health care for, xi-xii
ANC. *See* African National Congress
Anopheles mosquitoes, 24, 139–40

Lightning Source UK Ltd.
Milton Keynes UK
UKHW012021310821
389796UK00001B/89

9 780520 278219